# Isopropanol and Ketones

# in the Environment

# Isopropanol and Ketones in the Environment

Authors

**Samir Zakhari**
University of Pennsylvania
School of Medicine
Philadelphia

**Mitchell Leibowitz**
University of Pennsylvania
School of Medicine
Philadelphia

**Paul Levy**
University of Pennsylvania
School of Medicine
Philadelphia

**Domingo M. Aviado**
University of Pennsylvania
School of Medicine
Philadelphia

Editor-in-Chief

**Leon Golberg**

Chemical Industry Institute of Toxicology
Research Triangle Park, North Carolina

Solvents in the Environment Series
in
Uniscience Toxicology Series

**Published by**

CRC PRESS

CRC PRESS, Inc.
18901 Cranwood Parkway · Cleveland, Ohio 44128

**Library of Congress Cataloging in Publication Data**

Main entry under title:

Isopropanol and ketones in the environment.

Includes bibliographies and indexes.
1. Isopropyl alcohol – Toxicology. 2. Isopropyl alcohol – Physiological effect. 3. Butanone – Toxicology. 4. Butanone – Physiological effect. 5. Hexone – Toxicology. 6. Hexone – Physiological effect. I. Zakhari, Samir. II. Golberg, Leon.
RA1242.I8I86 615.9'51'31 76-41257
ISBN 0-8493-5197-9

This book represents information obtained from authentic and highly regarded sources. Reprinted material is quoted with permission, and sources are indicated. A wide variety of references are listed. Every reasonable effort has been made to give reliable data and information, but the author and the publisher cannot assume responsibility for the validity of all materials or for the consequences of their use.



International Standard Book Number 0-8493-5197-9

Library of Congress Card Number 76-41257
Printed in the United States

# THE AUTHORS

**Samir Zakhari** is an Instructor in Pharmacology at the University of Pennsylvania. He obtained his Ph.D. from the Slovak Academy of Science and is on leave of absence from the Pharmacology Unit, National Research Center of Cairo.

**Mitchell Leibowitz** received his Ph.D. from Columbia University in 1974 and is now an Instructor in the Department of Pharmacology of the University of Pennsylvania School of Medicine.

**Paul Levy** is a medical student and research fellow at the University of Pennsylvania School of Medicine. He received his B.A. from Rutgers University.

**Domingo M. Aviado** is the Professor of Pharmacology and Toxicology at the University of Pennsylvania School of Medicine at Philadelphia. He obtained his M.D. degree from the same University in 1948 and immediately joined the staff of the Department of Pharmacology.

This monograph is dedicated to Katharine R. Sturgis, M.D.

## FOREWORD

How often is one overcome by a sense of helplessness at the evil, the corruption, the degradation that surrounds us and permeates our society? We tend to overlook the nobility, the compassion, and the concern for others that mankind is capable of demonstrating. This volume is dedicated to one such person, a great lady whose life exemplifies these qualities of the human spirit, and whose activities in the field of preventive medicine constitute a superb record of service to her fellow men and women.

As one would judge from her many achievements, Dr. Katharine R. Sturgis is no ordinary person. Scholastically, scientifically, and clinically, Dr. Sturgis has distinguished herself and benefited the people of Philadelphia by her untiring efforts in the areas of pulmonary diseases, beginning with tuberculosis and moving on to lung cancer through the activities of the Philadelphia Pulmonary Neoplasm Research Project. Among the many successes of the Project, one may be mentioned particularly: the demonstration by Dr. Sturgis and Dr. William Weiss of the long-term respiratory effects of chloromethyl methyl ether and bis(chloromethyl) ether (*JAMA*, 234, 1139, 1975).

For many non-Philadelphians like myself, it was Dr. Sturgis' 11 years (1960–1970) as the first woman Chief Editor of the *AMA Archives of Environmental Health* that brought her to our attention. Her numerous editorials linger in one's memory for their deep philosophy, kindly and humorous spirit, and fine turn of phrase. Only those who have had to produce editorials on demand, year in and year out, can understand what is involved in this activity. Imagine superimposing such responsibility upon the duties of a wife and mother, the clinical rounds and service, the committees, and a host of other efforts in the community.

In view of the long association of Dr. Sturgis with lung disease and its prevention it is appropriate that her portrait should grace a volume in a series concerned with volatile solvents in the environment. In dedicating this book to her, the authors and I are happy to honor her and to wish her well.

Leon Golberg
Research Triangle Park, North Carolina

## ACKNOWLEDGMENT

The authors would like to thank David L. Snyder for his assistance in performing the experiments. The original investigation described in this monograph has been funded at least in part with funds from the United States Consumer Product Safety Commission under contract number CPSC-C-75-078. The content of this monograph does not necessarily reflect the views of the Commission, nor does mention of trade names, commercial products, or organizations imply endorsement by the Commission.

## OTHER TITLES IN THE CRC SOLVENTS IN THE ENVIRONMENT SERIES

Editor-in-Chief, **Leon Golberg, M.S., B.Chir., D.Sc., D.Phil., F.R.C.Path.**

METHYL CHLOROFORM and TRICHLOROETHYLENE in the ENVIRONMENT

By **Domingo M. Aviado, M.D., Samir Zakhari, Ph.D.,** both at the University of Pennsylvania School of Medicine; **Joseph A. Simaan, M.D.,** American University of Beirut; and **Andrew G. Ulsamer, Ph.D.,** United States Consumer Product Safety Commission.

NON-FLUORINATED PROPELLANTS and SOLVENTS for AEROSOLS

By **Domingo M. Aviado, M.D., Samir Zakhari, Ph.D.,** and **Tetsuya Watanabe, D.D.S.,** all at the University of Pennsylvania School of Medicine.

TABLE OF CONTENTS

# *Part 1*
# *Isopropanol*

Chapter 1

# REVIEW OF THE LITERATURE ON ISOPROPANOL

Unlike the fluorinated propellants[1] and the chlorinated solvents,[2] isopropanol is not generally recognized as being a frequent constituent of aerosol products. In actuality, it is found in a number of aerosol units including common household products as well as cosmetics and medicine.

Isopropanol was originally introduced into medicine as a substitute for ethanol (ethyl alcohol) about 50 years ago. Since isopropanol does not possess the inebriating properties of ethanol, it is less likely to be abused and is thus free from government regulations.

The chemistry and toxicology of isopropanol were reviewed by Hatch[3] in 1961. The only prior review known to us was written by Browning in 1952[4] and revised in 1965.[5] This monograph starts with an updating of the literature on isopropanol contained in these earlier reviews, with special reference to the inhalational toxicity of this solvent. This review of the literature is followed by reports of original investigation (Chapters 2, 3, and 4). The monograph concludes with a comparison of the hemodynamic effects of selected solvents and propellants in current use as aerosols (Chapter 5).

## A. Chemical and Physical Properties

Synonyms: Isopropanol; isopropyl alcohol; alcohol isopropylicum; secondary propyl alcohol; dimethylcarbinol; petrohol.

Chemical name: 2-propanol; ($CH_3CHOHCH_3$); C = 59.96%, H = 13.42%, O = 26.62%.

Molecular weight: 60.09.

Boiling point: 82.4°C; melting point −89.5°C; refractive index 1.3776 (20°C).

Specific gravity: 0.7809 at 25°C; 0.7855 at 20°C.

Vapor pressure: 33.0 mm Hg. at 20°C.

Vapor density: 2.07 (air = 1).

Volatility: Less than *n*-propyl alcohol; 21 times less than ether.

Solubility: Miscible in water, ethanol, ether, acetone, and benzene.

Description: Clear, colorless liquid with a slightly bitter taste and strong alcoholic odor.

Thermal stability: dehydrates at 260 to 400°C over alumina to form diisopropyl ether.

Flammability: Highly flammable; dangerous when exposed to heat or flame. Flash point (Cleveland open cup): 21°C. Lower explosive limit: 2.5%. Upper explosive limit: 5.2%. Autoignition temperature: 456°C.

Explosion hazard: Moderate, if exposed to flame.

Synthesis: The standard preparation of isopropanol involves the reaction between propylene and sulfuric acid to form isopropyl hydrogen sulfate, which is hydrolyzed to the solvent and sulfuric acid.

$$CH_3\ CH = CH_2 + H_2SO_4 \rightarrow (CH_3)_2CH\ O\ SO_3H \quad (1)$$

$$(CH_3)_2\ CH\ O\ SO_3H + H_2O \rightarrow CH_3\ CH\ O\ H\ CH_3 + H_2SO_4 \quad (2)$$

Reaction 1 is normally conducted in a packed absorption tower. Opposing streams of 85% sulfuric acid and propylene containing gas react exothermically at 21 to 27°C to produce isopropyl hydrogen sulfate at an approximate yield of 50%. The addition of three volumes of steam and consequent hydrolysis produce a dilute alcohol solution, which is further concentrated by rectification and purified with sodium hypochlorite or chlorine. Other synthetic methods include the hydration of propylene, the hydrogenation of acetone, furan, or carbon monoxide, and the fermentation of certain carbohydrates.

Determination of isopropanol: Methods have been reported for the determination of isopropanol in blood, tissue, and in pharmaceutical products. The quantitative microdetermination of the solvent in blood is accomplished by gas-liquid partition chromatography in a column of Porapak Q. Since blood may be analyzed in this manner without prior processing, prompt diagnosis of isopropanol intoxication is possible.[6] Simultaneous determination of isopropanol, ethanol, and methanol, as well as their metabolites, acetone and acetaldehyde, is possible in a single chromatographic column.[7] Other techniques for the chromatographic determination of isopropanol in blood, urine, and other biological fluids have been described.[8-10] Natelson and Stellate[11] designed an instrument for the identification of trace amounts of isopropanol in biological fluids, sewage water, and other aqueous solutions.

A tissue assay for isopropanol described by Schenkel and Schwartz[12] involves the determination of acetone, the oxidation product of isopropanol in oxidized tissue distillate. Falcone[13] employed a gas-solid chromatographic technique for the determination of isopropanol in drugs; he observed a 98.4 to 101.6% recovery range, with a mean of 100.2%.

### B. Medical Uses

Isopropanol enjoys widespread use as an antiseptic and disinfectant for home, hospital, and industrial use. It is thought to be approximately twice as potent as ethanol when applied for such purposes, but is free of the restrictive government regulations and potential for abuse of that solvent.[14,15]

Tanner and Wilson[16] showed that isopropanol is a more effective germicide than ethanol, but, like all alcohols tested in a series of 1 to 11 carbon lengths, is ineffective against microbial spores. Wantia[17] found isopropanol in a 50% benzine solution to be an effective germicide. Maximum germicidal potency is obtained with 50 to 80% water solutions; effectiveness is diminished at concentrations greater than 90%.[13] A 50% solution at 25°C kills *Staphylococcus aureus*, hemolytic streptococcus, *Escherichia coli* and *Bacillus typhosus* within 5 min after exposure. One hour exposures are inadequate against the spores of *Bacillus subtilis*.[18]

The antimicrobial potency of isopropanol is sufficient to allow its use in the aseptic preparation of needles, syringes, and surgical and laboratory instruments.[18] Though ineffective against spores, 50 to 80% isopropanol solutions are said to impart "clinical asepsis" in the sterilization of instruments not prone to harm by the alcohol.[15] Nancekievill[19] has described a simple, rapid, and inexpensive method for the decontamination of pulmonary ventilators, using a solution of 1% Savlon (chlorhexidine digluconate and cetrimide) in 70% isopropanol. One and a half pints of this solution satisfactorily decontaminates one ventilator.

A 50% isopropanol solution is thought to be as effective as a 70% ethanol solution for the decontamination of skin prior to parenteral injections, surgery, pre-surgical cleansing of hands, or the dressing of wounds.[14] The cleansing of skin with isopropanol solutions before needle puncture is of dubious value unless the solution is applied for at least 2 to 3 min, which is seldom the case.[15] As a skin disinfectant, isopropanol is most effective at concentrations less than 90%.

A solution of 85% isopropanol and 5% acetic acid has been shown effective in preventing acute bacterial infectious external otitis, or "swimming pool ear," when administered by droplet to the external ear canal after swimming. Such treatment is generally tolerated with little discomfort.[20]

Thirty percent solutions of the alcohol are useful in preparing medicinal liniments, tinctures of green soap, scalp tonics, and tincture of mercurophen.[12] Seventy percent isopropanol in water, "rubbing alcohol," is a popular massage solution for home and hospital use.

Since oral doses of isopropanol up to 5 ml are tolerated by humans, its use in oral pharmaceutical preparations has been suggested by Meyer and Klink.[21] Isopropanol is a constituent of various pharmaceutical preparations including local anesthetics, tincture of iodine, and bathing solutions for surgical sutures and dressings. It is prophylactic against burns caused by creosote, an infrequently used wound antiseptic.

### C. Commercial Uses

Because of its balance between alcoholic and hydrocarbon-like properties, isopropanol is a good solvent for a variety of oils, gums, waxes, resins, and alkaloids. Consequently, it is used in many industrial procedures and as a solvent in many consumer products, including drugs, cosmetics, cleaning compounds, paints, and lacquers. Hatch[3] tabulated over 200 uses of isopropanol.

#### *1. Industrial Uses*

Isopropanol is an economical substitute for ethyl alcohol in the preparation of nitrocellulose lacquers and shellacs used in the manufacture of capacitors, waterproof coatings, and insulators. As a solvent for assorted resins, it is useful in the preparation of cements, primers, varnishes, paints, and printing inks. It is a base ingredient of many shoe and hair-dye preparations. Its presence in household glues may produce rhinitis and bronchitis in abusers, or "sniffers," of these substances.[22]

#### *2. Cosmetics*

Isopropanol is widely employed in the preparation of many cosmetic products. As a solvent in lotions, perfumes, and shampoos, its acetone-like odor must be masked by the addition of fragrant synthetic agents or by raising the content of the perfume base.[23,24] The solvent is the main ingredient of certain sunscreen compounds, skin cleansers, and hair lotions. It is additionally found in nail polish solvents, make-up removers, deodorants, baby oils, and skin lotions. Seventy percent isopropanol in water, with perhaps perfume and coloring agents added, is a popular massage solution for home and hospital use.

*3. Cleansers*

Isopropanol's solvent properties are well applied in the preparation of industrial and home cleansing compounds, soaps, and detergents. Its versatility here derives from its usefulness as a "coupling agent" in homogenizing mixtures of water and organic solvents.[25] In combination with halogenated hydrocarbon solvents and emulsifying agents, it is useful in removing resins, sludges, grease, fat, and gums, and is an efficient rust preventative. With the addition of small quantities of synthetic detergent, 30 to 50% isopropanol in water solution is effective as a glass cleaner and polisher. A rapidly drying solvent, it is a popular base for stain removers and rug shampoos.[26,27] Mixtures of isopropanol, phosphoric acid, and *n*-butyl alcohol are useful for priming and polishing metal surfaces.

*4. Chemical Synthesis*

The crystallization, precipitation, and purification of various compounds in industry involve the use of isopropanol as a solvent. Assorted petrochemical processes involve treatment of reaction mixtures with isopropanol to remove impurities and reaction by-products. Its uses as a catalyst and as a chemical reaction medium have been catalogued by Hatch.[3]

*5. Dehydrating Agent*

Since it is miscible with water, isopropanol is an excellent dehydrating agent. It is thus employed in the processing of films, in metallurgical procedures, in drying laboratory apparatus, and in the preparation of biological specimens for histological and pathological study.

*6. Extraction of Natural Products*

Isopropanol is a useful solvent in the extraction of a variety of natural products including fat-soluble vitamins and antioxidants from vegetable and fish oils, and cottonseed and soybean oil from their respective plants, and for extracting fat-free fish protein.[28] Other extracts prepared with isopropanol include those of belladonna, cinchona, vanillin, ginger, pepper, pyrethrum, digitalis purpurea, sea moss, corn, peat, and wool wax.

*7. Miscellaneous Uses*

Other uses and proposed uses of isopropanol include: as a constituent of motor fuel, as a deicing fluid for the wings and propellers of aircraft and automobile windows, as a component of automotive antifreeze, as a corrosion inhibitor, and as a foam inhibitor.

**D. Aerosol Products**

Isopropanol is used as a solvent in aerosol products. Gleason et al.[29] have listed 68 aerosol products containing isopropanol which are as follows.

cleaners:
- window cleaners – four products (concentration of isopropanol known in two products: $<10\%$ and $\leq 10\%$)
- lens cleaner – two products (concentrations of isopropanol as follows: 36.5% and 8%)
- general purpose cleaner – one product (5.8% isopropanol)
- tape transport cleaner – one product

waxes and polishes:
- silver tarnish preventative – one product (86% isopropanol)

paints:
- paint – one product
- paint protector – one product (40% isopropanol)

automotive products:
- tire sealant – one product
- windshield deicer – three products (concentration of isopropanol known in one, 85%)
- general deicer – three products

specialty products:
- shoe and sock spray – two products (concentration of isopropanol known in one: $<25\%$)
- insect repellant – four products (concentra-

tion of isopropanol known in three: 8.7%, 25%, and 30%)

flea and tick spray – three products (concentration of isopropanol as follows: 14.7%, 16%, and 24%)

air fresheners – ten products (concentration of isopropanol known in six products: 3%, 9.6%, 10.8%, 12.3%, <20%, and 32%)

disinfectant – four products (concentrations of isopropanol known in three products: 12%, 17.8%, and 28.8%)

human medical and veterinary products:

foot fungicide – five products (concentration of isopropanol as follows: 25%, 47%, 50%, 59.5%, and 60%)

first aid spray – six products (concentrations of isopropanol known in five: 1.9%, 7%, 9.5%, 12%, and 12.2%)

skin soother – three products (concentrations of isopropanol: 1.5%, 10%, and 16%)

medicated vapor spray – one product (17% isopropanol)

veterinary pinkeye spray – two products (concentrations of isopropanol: 18% and 18%)

veterinary wound spray – nine products (concentrations of isopropanol: 10%, 18%, 18%, 26.25%, 37%, 52%, 60%, and 60.7%)

dehorning spray – one product (concentration of isopropanol: 57%)

Formulations containing isopropanol have been reported for the following aerosol products: hair sprays,[30-32] floor detergents,[33] shoe polishes,[34] fabric wrinkle remover,[35] insecticides,[36,37] burn ointment,[38] and skin refrigerant.[39]

## E. Human Investigation

There is no investigation directly relating to the health effects of the normal usage of aerosols containing isopropanol. There is also no documentation of the effects of the inhalation of isopropanol on human subjects. The reported toxicity in man is based largely on accidental ingestion of isopropanol, with a few cases following rectal administration and topical application to the skin.

### *1. Oral Ingestion*

Numerous case studies have been reported concerning the accidental ingestion of large quantities of isopropanol, usually in the form of "rubbing alcohol." Toxic doses of isopropanol may produce narcosis, anesthesia, coma, and finally death. These responses are not unlike those to large quantities of ethanol, but are observed at lower concentrations of the agent.

In most instances of severe intoxication, the odor of isopropanol or acetone is detected on the breath of the victim. Profound coma, often leading to death, has been reported after ingestion of one pint of "rubbing alcohol" (70% isopropanol).[40,41] This effect of the solvent is like that of ethanol, except that coma occurs at lower blood concentrations of the agent. In the cases reported, coma occurred at a blood isopropanol level of approximately 150 mg/100 ml. Survival periods ranged from 3 hr to 15 days. The victims manifested various degrees of hypothermia and shock. Acetonuria and widened T waves on the electrocardiogram were common diagnostic clues. Death occurred from paralysis of the central nervous system with no apparent anatomical damage. Hemoglobinuric nephrosis and hemorrhagic gastritis were often present.

Acute renal failure has been reported in several instances of consumption of 1 pint of "rubbing alcohol."[42,43] In two reported cases, renal failure was characterized by oliguria of 48 to 72 hr duration, followed by a prolonged diuresis of 8 to 12 days duration. Blood urea nitrogen levels rose to maximums of 120 mg/100 ml on the fourth day and 206 mg/100 ml on the fourteenth day in the two cases, before slowly returning to normal.

Wills et al.[44] administered small oral doses (2.6 and 6.4 mg/kg) of isopropanol daily for 6 weeks to human subjects. No significant deleterious effects were observed upon study of the chemistry and cellular composition of blood and urine, liver function, vision, and general state of health. Kemal[45] gave human subjects oral doses of the solvent and detected significant quantities of acetone, a metabolite of isopropanol, excreted in urine and expired air.

### *2. Rectal Administration*

Two reports of intoxication induced by rectal administration of isopropanol (100 ml and 450 ml) cite symptoms similar to those reported after ingestion of the solvent.[46,47] Deep stupor with depressed reflexes were observed initially; vomiting, facial flushing, acetonuria, and dizziness characterized the recovery periods.

*3. Topical Application*

Coma has been produced in hospital patients by topical application of isopropanol,[48-52] during "sponge baths" intended to reduce fever. Absorption through the skin in these cases is insignificant; patients readily inhale the vapor in poorly ventilated areas. Blood isopropanol concentrations after such treatments are reported as high as 10, 128, 200, and 220 mg/100 ml in several case studies; the latter three cases were marked by respiratory distress, severe stupor, and coma. Recovery in all instances was complete in 24 to 36 hr.

**F. Absorption, Metabolism, and Disposition in Animals**

Isopropanol is absorbed from all segments of the gastrointestinal tract, most rapidly in the small intestine and least rapidly in the stomach. Wax et al.[53] reported that absorption from isolated canine loops of intestine and stomach was 99% complete at the end of 2 hr, 82% being absorbed during the first 30 min. Intravenous injections of ethanol (0.8 cc/kg.) significantly reduced the absorption of isopropanol from the digestive tract.

Before 1940, the toxicity of isopropanol was compared to that of methanol, in which the long and cumulative action was attributed to slow rates of metabolism and excretion. But most data now point to a fairly rapid disposal of isopropanol, which is eliminated from the bloodstream of dogs within 24 hr after administration.

*1. Blood Levels*

The rate of disappearance of isopropanol from the bloodstream depends upon the efficiency of many organs, since high concentrations were found in urine, saliva, and stomach contents after intravenous injection in dogs, and since metabolites of isopropanol are excreted in urine and expired air.[54] Orskov[55] observed a linear decline in blood concentration of isopropanol and demonstrated that Widmark's $\beta$ constant varied from 0.0037 to 0.0058 for intraperitoneal doses ranging from 0.8 to 1.6 g/kg. Berggren[56] also demonstrated a linear decline of blood isopropanol concentration. Lehman et al.[57] obtained exponential curves for blood isopropanol concentrations, in contrast to the aforementioned reports and to the linear curves obtained by all for ethanol. Isopropanol is cleared from the bloodstream at approximately half the rate of ethanol.[58] A subcutaneous injection of insulin and dextrose reduced the rate of decline of blood alcohol by 50% after intravenous injection of isopropanol.[59]

Thirty minutes after ingestion of isopropanol (1.25 ml/kg) by dogs, concentrations of the compound in spinal fluid and blood were approximately equal, but higher concentrations were found in kidney and brain as opposed to liver and muscle.[53]

*2. Liver Metabolism*

Liver alcohol dehydrogenase (ADH) is apparently the major enzyme involved in the oxidation of isopropanol. A variety of alcohols, including ethanol are oxidized by this nonspecific enzyme. Pyrazole, a known inhibitor of ethanol oxidation, also blocks the in vivo metabolism of isopropanol in rats.[60] Dalziel and Dickinson[61] have shown that isopropanol is a poorer substrate for ADH than ethanol or *n*-propanol. Since some workers contend that isopropanol oxidation involves a different enzyme, the nature of the in vivo oxidation is still unsettled.

The elimination of isopropanol from the blood of rats is, however, inhibited by *n*-propanol and ethanol; conversely, isopropanol inhibits the elimination of ethanol.[61-63] A competitive substrate inhibition of ADH is thus indicated.

Acetone produced by ADH in the liver is eliminated from the human body in urine and expired air. Acetone is apparent in the urine within the first hour after administration and may still be detected after 48 hr. Acetone appears on the breath within 15 min. The acetone found in the urine normally represents a small fraction of the total ingested isopropanol, but could be increased up to 40% of the total by liberal consumption of fluids.[45] Some estimate that 70 to 90% of ingested isopropanol is metabolized to acetone.[64,65]

Modifications of the liver extramitochondrial NAD/NADH ratio produced by ethanol are not seen with isopropanol administration in rats. This may be explained by the further metabolism of acetone, which is transformed in the liver into acetol.[66] The acetone is reduced to 1, 2-propanediol in a reaction that involves the reoxidation of NADH produced in the ADH oxidation of isopropanol. Thus the extramitochondrial redox state is balanced.[58]

*3. Urinary Excretion*

Isopropanol is not concentrated in the urine. After filtration by the glomeruli it is reabsorbed along with water from the renal tubules. Since the urine concentration comes into equilibrium with the blood concentration, isopropanol excretion is increased only by increasing urine output. There is no fixed relationship between the blood and urine isopropanol concentrations, like that for ethanol. Gaillard and Derache[67] recovered only small quantities of ingested isopropanol in rat urine. They detected 0.34% of the ingested compound, compared to 2.02% of ingested ethanol.

**G. Acute Toxicologic Investigation in Animals**

The comparisons of the lethal dosages in animals indicate consistently that isopropanol is approximately 2 times more toxic than ethanol (see references cited by Morris and Lightbody[68]). This review covers the lethal dosages for isopropanol only.

*1. Oral Administration*

Lehman and Chase[69] administered 50% isopropanol solutions via stomach tube and reported the following $LD_{50}$: rats, 6.73 ml/kg; rabbits, 6.41 ml/kg; and dogs, 6.15 ml/kg. Other investigators have reported oral lethal dosages of 5 to 10 ml/kg in dogs,[70] cats,[70,71] rabbits,[72] and rats.[73] In the last-mentioned species, three age groups were tested by Kimura et al.[74] Their results were as follows: the oral $LD_{50}$ for 14-day-old rats was 5.6 ml/kg, which was lower than the young adults (6.0 ml/kg) and the older rats (6.8 ml/kg). The cause for the susceptibility to the toxic effect of isopropanol of the immature animals has not been elucidated.

*2. Intravenous Injection*

Lehman and Chase [69] determined the doses that produced anesthesia and death in rabbits and dogs:

| | Anesthetic dose | Lethal dose |
|---|---|---|
| Rabbits | 3.23 ml/kg | 8.23 ml/kg |
| Dogs | 3.35 ml/kg | 5.12 ml/kg |

Macht[70] reported the intravenous lethal dose in cats to be 2.5 ml/kg. On the basis of these results, cats are more sensitive than dogs and rabbits to the lethal action of isopropanol.

*3. Inhalational Toxicity*

Rats exposed to 12,000 ppm (2.94 mg/l) isopropanol-in-air mixtures survived 4-hr exposures, though 8-hr exposures killed half the animals.

Dogs and rats exposed to air saturated with isopropanol vapor for 1 hr daily and for 4 to 7 days show mild and transient incoordination after exposure.[69] Air saturated with isopropanol produced a slight unsteadiness of gait in dogs and rabbits after 1 hr of exposure. A similar response was noted in rats after 15 to 20 min of inhalation.[71]

**H. Subacute Toxicologic Investigation in Animals**

Acquired tolerance to isopropanol has been reported in a variety of species. Morris and Lightbody[68] noted "that the ingestion of the substance produces a form of intoxication, especially in the early period of the test is apparent, the violence and duration of the same depending on the species. It is clear, however, that in most instances, barring perhaps cats, a tolerance is quickly established, and thereafter the appearance of the intoxification differs in no respect from that produced by ethyl alcohol."

*1. Rats*

The subacute toxicity of isopropanol was investigated in white rats given isopropanol solution (0.5 to 10% in drinking water) ad lib. During 27 weeks of observation, growth and weight gain were retarded in male rats during the first 13 weeks but accelerated during the final 14 weeks. Increased caloric intake was thought to account for the increased weight gain compared to controls. Growth and body weight in female rats were depressed approximately 10% compared to control, though appetite in these rats was judged normal. Examination of the brain, heart, lung, liver, kidneys, spleen, pituitary, and adrenals showed no gross or microscopic changes.[69]

Boughton[76] provided groups of rats with 5% ethanol and 5% isopropanol solutions as their water supply for 9 months. Average fluid consumption during that time was depressed 41% in the isopropanol group, compared to controls, and increased 48% after withdrawal of the alcohol. Corresponding figures for the ethanol group were a

36% depression followed by a 37% increase. Weight depression, "noisy breathing," and sluggishness during maze testing were pronounced in both groups during the experimental period. Death rates were near normal in all groups.

*2. Guinea Pigs*

Isopropanol was shown to be a more potent narcotic than ethanol, as measured by the induction of paralysis in guinea pigs, when administered by stomach tube.[77]

*3. Dogs*

Lehman and Chase[69] observed that "prolonged imbibition of large amounts of isopropanol for a period of seven months renders dogs tolerant to the alcohol, as manifested by a greater degree of neuromuscular coordination at similar blood alcohol concentrations after habituation than before, and by increased elimination of isopropanol." No secondary toxic symptoms are manifested in dogs after long-term ingestion of large amounts of isopropanol daily. Since no cumulative effects are observed, chronic poisoning is essentially no different than acute toxicity.

**I. Pharmacodynamic Effects**

In the toxicologic investigation of isopropanol discussed in the two preceding sections (G and H), the most conspicuous effect is depression of the central nervous system which is seen consistently with most alcohols.[78-81] The details of the effects on the central nervous system and other organs are discussed in this section.

*1. General Anesthesia*

The use of isopropanol as a laboratory anesthetic for cats was first suggested by Macht.[82] The alcohol is administered via a stomach tube, after the animal is first anesthetized with ether, in doses of 5 to 5 1/2 ml/kg, diluted two to three times with water. With such doses, satisfactory anesthesia, circulation, and respiration were maintained for hours. Intraperitoneal administration of 1.6 g/kg of the solvent produces partial anesthesia in rabbits.[55] Seeman[83] studied a series of alcohols, including isopropanol, and correlated alcohol-induced erythrocyte membrane stabilization with alcoholic blockade of frog sciatic nerve and suppression of the tadpole reflex. He concluded that there was a positive correlation of effects on the erythrocyte and the depression of nerve conduction.

*2. Depression or Stimulation of Other Parts of the Nervous System*

Isopropanol has narcotic effects on the central nervous system, with a potency approximately twice that of ethanol. However, its mawkish taste and inability to produce exhilaration or euphoria provide no basis for its sporadic use as a beverage.

Administration to human subjects produced dizziness within 30 min, followed 2 to 3 hr later by headache and mental depression. The headache is variably characterized as moderate to very severe and, in some instances, lasted longer than 24 hr. Muscular incoordination and mental disturbance were not conspicuous. The CNS depression was apparent at doses considerably below the estimated lethal dose.[68] In an isopropanol intoxication case brought to autopsy, Glasgow and Ferris[84] observed a diffusely edematous cerebral cortex, with no evidence of encephalitis or meningitis.

Animal studies similarly demonstrate a pronounced intoxication after administration of isopropanol by oral, intraperitoneal, or inhalation routes. Hind leg paralysis, unsteadiness, generalized ataxia, respiratory depression, and stupor were commonly observed symptoms.[71] Cupping of the optic disc, with no detectable blindness, was noted in rabbits.[85]

While the intoxicating effects of isopropanol are well documented, their biochemical basis is poorly understood. Several investigators report a marked depression of respiration in rat brain cortex slices in perfusion with small concentrations of isopropanol.[86,87] The oxidation of glucose-$C^{14}$ to $C^{14}O_2$ is depressed by about 12%. Others have shown that isopropanol inhibits potassium uptake in rat cortex slices and depresses sodium-, potassium-, and magnesium-dependent adenosine triphosphatase activity in rat brain microsomes.[88] These results suggest that the depressant action of isopropanol on the central nervous system may result from an inhibition of oxidative metabolism and membrane cation transport in neurons.

Isopropanol inhibits the activity of peripheral neurons as well. By depressing the resting membrane potential, sodium conductance, and to a lesser extent, potassium conductance of a giant squid axon, the solvent produces a fall in the magnitude of the action potential peak.[89] At concentrations greater than 351 mMol/liter, isopropanol blocks nervous conduction in a frog sciatic nerve to a degree that correlates well with the solvent's propensity to penetrate a monomolecular nerve lipid layer.[90] X-Ray diffraction and electron microscope studies reveal pronounced structural disorganization of myelin sheaths and removal of myelin from rat sciatic nerve sheaths[91] after exposure to isopropanol. The alcohol's intermediate hydrophilic and lipophilic properties are thought to promote the disruption of neural lipoproteins.

Enhancement by isopropanol of transmitter release at neuromuscular junctions has been reported.[92] In a single case, cerebrospinal fluid protein in a 2-year-old child was markedly elevated (200 mg/100 ml) following ingestion of the alcohol.[93]

Isopropanol has been shown to be an effective anticonvulsant against clonic convulsions when given gastrically to rats, cats, and rabbits.[94,95] The alcohol and its immediate metabolite, acetone, are thought to be responsible for this effect. Since isopropanol raises the threshold for electrical convulsions and potentiates the response to diphenylhydantion in rats, it has been suggested that the isopropyl group may be useful in improving current antiepileptic agents or in developing new ones.[96] The inhalation of isopropanol in rats causes excitation of the olfactory pathways.[97,98]

*3. Cardiovascular Toxicity*

Isopropanol is apparently not toxic to the heart and the circulatory system unless administered in very large doses. Ingestion of 25 ml in 100 ml of water by humans produced hypotension, facial flushing, bradycardia, and dizziness. Hypotension seems to be the most serious circulatory effect of isopropanol intoxication.[42] Widened T waves on electrocardiogram are an important diagnostic clue to isopropanol intoxication.

The solvent produces a concentration-dependent reduction in the contractile force generated by an isolated guinea pig ventricular strip. A concentration of 0.3 *M* isopropanol causes a 60% reduction of the peak myocardial contractile tension generated.[99] In their studies of isolated frog hearts, Ringer and Sainsbury found isopropanol to be more than twice as toxic as ethanol in depressing cardiac activity.[100] Topical application of the solvent on guinea pig skin fails to produce increased permeability and erythema of superficial blood vessels.[101]

*4. Pneumotoxicity*

In addition to its anesthetic and narcotic actions, isopropanol tends to paralyze the respiratory system. The magnitude of this effect is approximately twice as great as that of *n*-propanol or ethanol. Because the central nervous system is highly vulnerable to this solvent, a person suffering from isopropanol intoxication is likely to show deep refractory narcosis, areflexia and depressed respirations.[43] Paralysis and dyspnea may portend death. Death following ingestion of isopropanol is usually due to paralysis of the respiratory center.[42] Aspiration of 0.2 ml isopropanol by rats produced death from respiratory failure within 1 to 5 hr.[73,102]

*5. Hepatotoxicity*

It has been observed that acute ethanol administration increases hepatic triglyceride content as a result of enhanced lipogenesis and inhibition of fatty acid breakdown. Large increases of triglyceride content are also observed in rat livers 4 hr after gastric administration of a large (6 g/kg) dose of isopropanol.[103] Since it has been shown that isopropanol administration increases the free fatty acid level in blood, causes no additional hepatic fatty acid synthesis, does not alter the rate of fatty acid release from the liver, and leads to new hepatic triglyceride with the same fatty acid distribution as found in adipose tissue, it is proposed that isopropanol accelerates the mobilization of fatty acids from adipose tissue and increases hepatic triglyceride synthesis.[103-106]

While increasing the incorporation of fatty acid into hepatic triglyceride, isopropanol may direct fatty acid utilization away from hepatic phospholipid synthesis.[104,107,108] In other studies, isopropanol was shown to have produced no effect on hepatic cholesterol and phospholipid levels 17 hr following oral administration of 6 g/kg doses to rats.[109,110] Blood lipid levels also were not changed significantly.

Pretreatment of animals with isopropanol (2.5 mg/kg orally) has been shown to enhance the acute toxicity of carbon tetrachloride, as measured by (1) the elevation of serum glutamic-pyruvic transaminase levels, (2) the elevation of hepatic triglyceride content, (3) the depression of hepatic glucose-6-phosphatase activity, and (4) the elevation of hepatic vascular resistance.[111-113] A 2.5 ml/kg dose of isopropanol followed by a 0.1 ml/kg dose of carbon tetrachloride has greater toxicity in rats than carbon tetrachloride administered alone.[114-116] In addition, only carbon tetrachloride with isopropanol produced hyperbilirubinemia.

The potentiation of the carbon tetrachloride response by isopropanol is most pronounced when the pretreatment precedes the administration by 16 to 18 hr. No enhancement is observed when 2 hr separate the two.[117] It appears that acetone, an immediate metabolite of isopropanol, is responsible for enhancement of carbon tetrachloride toxicity – hence, the requirement for a delay after pretreatment.[118] Pretreatment with acetone causes a more immediate enhanced carbon tetrachloride hepatotoxicity.

Isopropanol, or acetone, is thought to enhance carbon tetrachloride hepatotoxicity by activating the microsomal drug-metabolizing enzyme system which converts carbon tetrachloride to an active metabolite.[111] The capacity of rat liver microsomes to bind carbon tetrachloride is enhanced by either isopropanol or acetone.[118] Aminotriazole, an inhibitor of microsomal enzymes, reduces the enhancing capability of both isopropanol and acetone.[112] Pyrazole, in addition to slowing the rate of clearance of isopropanol from the bloodstream of rats, reduces by 40% the enhancement of carbon tetrachloride hepatotoxicity by isopropanol.

The oxidative deamination of 5-hydroxytryptamine (serotonin) to produce 5-hydroxyindoleacetic acid in perfused rat and frog livers is depressed by isopropanol.[119-122] The simultaneous increased production of 5-hydroxytryptamines-O-glucuronide and 5-hydroxytryptamine-O-sulfate may be an alternative metabolic pathway when deamination is blocked. A reduction of the NAD/NADH ratio may depress the rate of deamination.

Forsander[123] observed that isopropanol decreased carbon dioxide production in rat liver slices, exerted no influence on hepatic oxygen uptake, and had no significant effect on hepatic lactate/pyruvate ratio. The solvent also depresses chloral hydrate metabolism in rat liver slices,[124] inhibits the oxidation of methanol to formaldehyde in liver homogenates,[125] and severely inhibits hepatic D-$\beta$-hydroxybutyrate dehydrogenase.[126] Intravenous administration of isopropanol (2 g/kg) increased the concentration of adenosine triphosphate and decreased that of adenosine diphosphate in rat liver after 5 min.[127]

*6. Nephrotoxicity*

Persual of the literature reveals several reports of nephrotoxic activity of isopropanol. Chapin[42] cited acute renal insufficiency as a complication of isopropanol poisoning in humans. The insufficiency state or "lower nephron nephrosis" was thought to result from both a direct action of isopropanol on renal tissue and from renal anoxia in shock secondary to isopropanol poisoning. The normal concomitants of renal insufficiency, including anuria followed by oliguria, edema, and nitrogen retention were observed.

Subcutaneous injections of 5 to 10 ml/kg isopropanol produced hematuria in mice, rats, and rabbits according to Levy-Hochman.[128] The site of the hematuria was identified as the renal medullary intertubular spaces, but the nature of the hematuric process was not determined.

Leng[129] has demonstrated a significant diuretic response to intravenous injection of isopropanol in rats. The diuretic action of a 0.10 ml/kg dose was observed after a 10-minute latent period and was 36% that of a comparable dose of ethanol.

Lehman et al.[54] administered intravenously near-lethal doses (3.8 ml/kg) of isopropanol to dogs. The dogs were sacrificed 36 hr later and the kidneys were examined for pathological changes. Necrosis of some tubular epithelium and a decrease in the number of nephrons were observed.

*7. Oral and Gastrointestinal Toxicity*

Baer and Archard[130] reported a case of injury to the gums following incorrect use of isopropanol as a mouth rinse. On the gastric mucosa, this solvent causes inflammation, bleeding, and inhibition of pepsin.[131,132]

*8. Embryo- and Fetotoxicity*

A group of male and female rats were placed, at the age of 40 days, on a diet that included 2.5% isopropanol in their drinking water. The rats were mated when they were 120 days old and the process was repeated through two generations. No significant differences were observed between the growth rates of the second generation alcoholized rats and second generation control rats. The ingested solvent produced no harmful effects on the reproductive activity and embryonic development of the rats.[54]

*9. Carcinogenicity*

The observation that workers involved in the production of isopropanol suffered an above-normal incidence of respiratory tract tumors prompted a search for a carcinogen in the synthetic process.[133,134] Malignant tumors of the paranasal sinuses accounted for a disproportionate number of the observed cases, some fatal.[135]

Isopropanol was found "nontumorigenic via the inhalation, injection, and skin-painting routes."[135] It is thus suggested that a waste product of the synthetic process, perhaps isopropyl oil, is a carcinogen. Alternatively, isopropanol in combination with isopropyl oil or fusel oil may be carcinogenic.[136] Isopropanol alone is generally thought to have no carcinogenic activity.

*10. Dermatotoxicity*

The dermatotoxic potential of isopropanol is small as evidenced by its inclusion in shampoos, skin creams, and common "rubbing alcohol." Fregert et al.[137] and McInnes[138] report several cases of hypersensitivity to isopropanol, characterized by delayed, severe eczematous reactions after dermal application of the compound. Daily application of 50% isopropanol to the faces of albino rats for 187 days produced no apparent injury to skin.[76] Macht sponged isopropanol on a 16-$cm^2$ area of shaved dog skin during a 2-week period; no local or systemic symptoms were noted.[70] However, when isopropanol is administered to the blood supply of the skin, there is an increase in capillary permeability.[140,141] The absorption of isopropanol through human skin was studied by Blank,[139] who observed that absorption was enhanced by increasing concentration of the compound in the presenting vehicle and decreasing the solubility of the compound in the vehicle.

## J. Clinical Toxicology

The symptomatology of isopropanol poisoning was summarized by Von Oettingen[142] as follows:

> Dizziness, muscular and nervous disturbances have been reported; also bradycardia, and lowering of the blood pressure. Ingested isopropyl alcohol becomes concentrated in certain organs, particularly in the brain, heart muscle, and kidneys. Pathologically, moderate reversible fatty infiltration may be found in these organs. Large doses cause ataxia and narcosis; paralysis and dyspnea precede death.

Others have shown that large oral doses of isopropanol may produce flushing of the face, irritation of gastric mucosa, vomiting, depression, headache, and unconsciousness.[54,68,143,144] An estimated minimum fatal dose of isopropanol is 166 ml, ingested rapidly on an empty stomach. Others report the lethal dose as 160 to 240 ml.[65] Death often occurs in 24 to 36 hr from respiratory paralysis.

While no specific therapy is indicated in cases of isopropanol intoxication, Adelson[40] advises that treatment should include "immediate evacuation of the stomach with continuous gastric suction until the victim emerges from coma, antishock measures, oxygen via an endotracheal tube, maintenance of fluid balance, correction of electrolyte disturbances, and symptomatic care generally."

Because physiological mechanisms for metabolizing and excreting isopropanol are slow, hemodialysis has been successfully employed in treating isopropanol poisoning.[145] King et al.[146] describe their success in using hemodialysis to treat a patient who had ingested 1 pint of "rubbing alcohol" (70% isopropanol). On admission, 45 min after ingestion, the patient was comatose, respiration was shallow at 14/min, and blood pressure was 60/40. After 2 hr of dialysis, blood pressure was 120/70; the patient was conscious and agitated. After 5 hr, blood pressure was 140/70; the patient was quiet and responsive. The rest of the recovery was uneventful.

In instances when hemodialysis is unavailable, the less efficient and more time consuming peritoneal dialysis has been successfully employed. In a reported case[147] dialysis was conducted through 20 cycles to treat a patient after ingestion of 1 pint of rubbing alcohol. Isopropanol was undectable in blood after just eight cycles.

## K. Threshold Limit Value (TLV)

In 1963, the American Conference of Governmental Industrial Hygienists established the value of 400 ppm (0.098 mg/l) as the threshold limit value for isopropanol, and this is the TLV currently enforced in the United States.[148] This level was noted by Nelson et al.[149] to cause mild irritation of the eyes, nose, and throat, whereas, twice the TLV (800 ppm or 0.196 mg/l) intensified the symptoms. There are no other observations which compel a change in the current TLV. In Russia, a maximum allowable concentration of 2.4 ppm

(0.006 mg/l) is recommended but the basis for this unusually low level is not explained by Baikov et al.[150] and Gorlova.[151]

## L. Conversion Table for Concentrations of Isopropanol

This review of the literature concludes with a table that is useful in converting concentrations of isopropanol on a volume/volume basis to weight/volume. The following relationships are based on the constants for isopropanol: molecular weight of 60.09; volume occupied by one molecular weight at 25.9°C = 24.58 liters; therefore, 2.45 g of isopropanol occupies 1 liter of gas.

| mg/l | ppm | v/v % |
|---|---|---|
| 0.000245 | 1 | 0.0001 |
| 0.0245 | 100 | 0.01 |
| 0.1 | 408 | 0.04 |
| 0.245 | 1000 | 0.1 |
| 1.0 | 4082 | 0.41 |
| 2.45 | 10,000 | 1 |
| 10.0 | 40,816 | 4.1 |
| 12.25 | 50,000 | 5 |
| 15.0 | 61,224 | 6.1 |
| 24.5 | 100,000 | 10 |
| 25.0 | 102,040 | 10.2 |
| 49.0 | 200,000 | 20 |

## Chapter 2
## ACUTE INHALATIONAL, ORAL, INTRAPERITONEAL, AND SUBCUTANEOUS TOXICITY IN MICE AND RATS

In the comparison of lethal dosages of solvents, the mouse has been selected as the reference animal. So far, the two chlorinated hydrocarbons have the following lethal dosages as previously reported from this laboratory.[2]

| | Methyl chloroform | Trichloroethylene |
|---|---|---|
| Oral $LD_{50}$ | 14.08 ± 2.05 g/kg | 2.85 ± 0.55 g/kg |
| Intraperitoneal $LD_{50}$ | 2.34 ± 0.32 g/kg | 1.20 ± 0.31 g/kg |
| 5-min inhalational $LC_{50}$ | 0.81 ± 0.14 g/l | > 0.535 g/l |
| 20-min inhalational $LC_{50}$ | 0.27 ± 0.08 g/l | 0.22 ± 0.09 g/l |

The above results indicate that trichloroethylene has lower lethal dosages and concentrations than methyl chloroform. The administration of either chlorinated solvent caused changes in the electrocardiogram which indicated cardiotoxicity.

The first non-halogenated solvent examined in this laboratory is isopropanol which has been investigated to a limited extent. There are no reported measurements of $LD_{50}$ or $LC_{50}$ for isopropanol. The only account is by Weese[152] who, in 1928, reported the lowest oral lethal dose of 0.192 g/kg in mice. This value, which has not been confirmed, is considerably lower than the oral $LD_{50}$ reported by Lehman and Chase[69] of 6 to 7 ml/kg (4.7 to 5.5 g/kg) in rats, rabbits, and dogs.

The experiments reported below were designed to compare the acute toxicity of isopropanol in mice and rats. The experiments on mice are for comparison with halogenated solvents, whereas the additional experiments on rats are preparatory to the investigation of physiological changes in the lungs and heart reported in Chapter 3.

### A. Methods

Toxicity studies were performed on unanaesthetized male mice of the Swiss-Webster strain, weighing from 21 to 25 g (average 22.5 g), and on male Wistar albino rats (Charles River Breeding Laboratories, Wilmington, Massachusetts) weighing from 200 to 250 (average 219 g). Isopropanol (Certified A.C.S., Fisher Scientific Company) was administered to experimental animals via various routes of administration.

#### *1. Oral Route*

Mice were divided into five groups with ten animals in each group. On the basis of previous determination of the toxic dose, each group of mice received various dosage levels of isopropanol administered via gastric tubing (2, 4, 6, 8, and 10 g/kg). Rats were divided into seven groups, each of five rats, to which the following doses of isopropanol were administered: 3.1, 3.5, 3.9, 4.3, 4.7, 5.1, and 5.5 g/kg.

#### *2. Intraperitoneal Route*

Five groups of ten mice each were injected with the following dosage levels of isopropanol: 0.5, 1, 2, 3, and 4 g/kg. Rats were divided into six groups of five rats, to which the following doses of isopropanol were administered: 1.2, 1.6, 1.8, 2.0, 2.1, and 2.3 g/kg.

#### *3. Inhalational Route*

Mice received isopropanol vapors by inhalation for a period of 2 hr. Various groups of mice were introduced into a 10-liter glass chamber with an inlet and outlet for the gas and were exposed to 2.5, 12.3, 24.5, 36.8, and 49.0 mg/l (10,000 50,000, 100,000, 150,000, and 200,000 ppm) concentrations of isopropanol. Vapor mixtures were allowed to circulate in the inhalational chamber at a rate of 2 to 3 l/min. Various concentrations were prepared by volatilizing the calculated volume of isopropanol into a known volume of air at standard pressure and 25°C, as previously described.[2]

#### *4. Subcutaneous Route*

Each of six groups of rats, five per group, received one of the following six doses of isopropanol; 5.1, 5.5, 5.7, 5.9, 6.1, and 6.3 g/kg.

Mice and rats were allowed free access to food

and water, and the percentage mortality was determined after 24 hr. $LD_{50}$ and $LC_{50}$ were calculated according to the probit method.[153]

## B. Results

Mice and rats showed signs of depression of the central nervous system following the administration of lethal doses or concentrations of isopropanol by all the routes, i.e., oral, intraperitoneal, subcutaneous, and inhalation. The results on the mice are summarized in Tables 2.1 to 2.3 and those on the rats in Tables 2.4 to 2.6. The lethal dosages for the two animal species are as follows:

| | Isopropanol | |
|---|---|---|
| | Mice | Rats |
| Oral $LD_{50}$ | 4.8 ± 1.80 g/kg | 4.70 ± 0.78 g/kg |
| Intraperitoneal $LD^{50}$ | 1.28 ± 0.61 g/kg | 1.87 ± 0.23 g/kg |
| Subcutaneous $LD_{50}$ | | 5.70 ± 0.23 g/kg |
| 120-min. inhalational $LC_{50}$ | 0.01039 ± 0.00368 g/l | |

TABLE 2.1

**Mortality Rate and $LD_{50}$ (24 hr) of Isopropanol, Administered Orally to Mice**

| Groups | Dose (g/kg) | % Death | $LD_{50}$ (g/kg) 24 hr | Regression coefficient |
|---|---|---|---|---|
| 1 | 2 | 20 | | |
| 2 | 4 | 30 | | |
| 3 | 6 | 60 | 4.8 ± 1.8 | 0.9839 |
| 4 | 8 | 90 | | |
| 5 | 10 | 100 | | |

TABLE 2.2

**Mortality Rate and $LD_{50}$ (24 hr) of Isopropanol, Injected Intraperitoneally to Mice**

| Groups | Dose (g/kg) | % Death | $LD_{50}$ (g/kg) 24 hr | Regression coefficient |
|---|---|---|---|---|
| 1 | 0.5 | 10 | | |
| 2 | 1 | 20 | | |
| 3 | 2 | 80 | 1.28 ± 0.61 | 0.9766 |
| 4 | 3 | 100 | | |
| 5 | 4 | 100 | | |

TABLE 2.3

**Mortality Rate and $LC_{50}$ (24 hr) of Isopropanol, Administered by Inhalation for 120 min to Mice**

| Groups | Concentration (mg/l) | % Death | $LC_{50}$ (mg/l) 24 hr | Regression coefficient |
|---|---|---|---|---|
| 1 | 2.45 | 0 | | |
| 2 | 12.25 | 60 | | |
| 3 | 24.50 | 70 | 10.39 ± 3.68 | 0.9899 |
| 4 | 36.75 | 90 | | |
| 5 | 49.00 | 100 | | |

TABLE 2.4

**Mortality Rate and $LD_{50}$ (24 hr) of Isopropanol, Administered Orally to Rats**

| Groups | Dose (g/kg) | % Death | $LD_{50}$ (g/kg) 24 hr | Regression coefficient |
|---|---|---|---|---|
| 1 | 3.1 | 0 | | |
| 2 | 3.5 | 10 | | |
| 3 | 3.9 | 20 | | |
| 4 | 4.3 | 20 | 4.70 ± 0.78 | 0.9452 |
| 5 | 4.7 | 60 | | |
| 6 | 5.1 | 60 | | |
| 7 | 5.5 | 100 | | |

TABLE 2.5

**Mortality Rate and $LD_{50}$ (24 hr) of Isopropanol, Administered by Intraperitoneal Injection to Rats**

| Group | Dose (g/kg) | % Death | $LD_{50}$ (g/kg) 24 hr | Regression coefficient |
|---|---|---|---|---|
| 1 | 1.2 | 0 | | |
| 2 | 1.6 | 20 | | |
| 3 | 1.8 | 20 | 1.87 ± 0.23 | 0.9394 |
| 4 | 2.0 | 60 | | |
| 5 | 2.1 | 80 | | |
| 6 | 2.3 | 100 | | |

TABLE 2.6

**Mortality Rate and $LD_{50}$ (24 hr) of Isopropanol, Administered by Subcutaneous Injection to Rats**

| Groups | Dose (g/kg) | % Death | $LD_{50}$ (g/kg) 24 hr | Regression coefficient |
|---|---|---|---|---|
| 1 | 5.1 | 0 | | |
| 2 | 5.5 | 20 | | |
| 3 | 5.7 | 40 | 5.70 ± 0.23 | 0.9623 |
| 4 | 5.9 | 80 | | |
| 5 | 6.1 | 80 | | |
| 6 | 6.3 | 100 | | |

## C. Discussion

The above results represent the first reported attempt to determine the lethal concentration of isopropanol in animals. The highest concentration tested was 49 mg/l (200,000 ppm), which was lethal to all mice after 120 min of exposure. There is no question that isopropanol vapor by inhalation is less toxic than the halogenated hydrocarbons, methyl chloroform and trichlorethylene, which in concentrations of 0.27 g/l and 0.22 g/l, respectively, are lethal in 20 min It is not possible to administer by inhalation isopropanol vapor in quantities that are lethal to the mouse within less than 1 hr.

Contrary to the implication from the published literature that mice are more sensitive than rats to the lethality of isopropanol, the above results indicate that there is no difference between the two animal species. The oral $LD_{50}$ as well as the intraperitoneal $LD_{50}$ are about equal in the two species. That the oral $LD_{50}$ is three to four times greater than the intraperitoneal $LD_{50}$ is an indication of the differnece in rate of absorption from the gastrointestinal tract and from the peritoneal cavity. That the oral $LD_{50}$ is about equal to the subcutaneous $LD_{50}$ in rats excludes the importance of hepatic circulation in influencing lethality. If the oral $LD_{50}$ were larger than the subcutaneous $LD_{50}$, there would be the possibility that the liver is metabolizing isopropanol as rapidly as it is absorbed. The results indicate that isopropanol administered by any route can reach the cardiovascular and the central nervous system. The loss of consciousness, which represents depression of the central nervous system, precedes death. The effects on the circulatory system are elucidated in the next chapter.

## D. Summary

Isopropanol vapor up to the point of maximum saturation in air is not lethal to mice in less than 1 hr. The $LC_{50}$ administered for 120 min is 10.39 ± 3.68 mg/l (49,120 ppm) in mice. The oral and intraperitoneal $LD_{50}$s of isopropanol are about equal in mice and rats.

# Chapter 3
# PHARMACODYNAMIC EFFECTS OF ISOPROPANOL ON THE RESPIRATORY AND CIRCULATORY SYSTEMS IN RATS

In the review of the literature on isopropanol (see Chapter 1), there was limited information relating to the respiratory and circulatory systems. There are reports of intoxication from ingested isopropanol characterized by hypotension and respiratory depression.[42,43] Death is believed to be caused by paralysis of the respiratory center which is brought about by the narcotic effect of isopropanol in particular and of alcohols in general. There is no information as to the status of the heart and it is not known if the hypotension is caused by cardiac depression or by vasodilation or by both. Depression of cardiac contractility has been reported in the isolated frog heart[100] as well as in the isolated guinea pig ventricular muscle strip so that it is not possible to exclude cardiac failure as a cause of death.

The experiments reported in this chapter represent the first attempt to examine the pharmacodynamic effects of isopropanol in one animal species. The rat has been used for the first time in the comparative study of solvents to supplement data on the mouse and dog reported in earlier monographs of this series.[2]

## A. Methods

A total of 120 male Wistar rats were used in this investigation. The rats were divided by weight into two groups of 200 to 250 g and 251 to 300 g. All of them were used in an anesthetized state after intraperitoneal injection of 65 mg/kg sodium pentobarbital. The 200- to 250-g rats were prepared for measurements of pulmonary function; the 251- to 300-g rats were prepared for measurement of pulmonary function, arterial pressure, and for recording of the Lead II electrocardiogram. The details of each major procedure are as follows:

### *1. Tracheal Cannulation*

The rat was anesthetized with sodium pentobarbital and the trachea was exposed by a midline neck incision. The trachea was cut and a polyethylene cannula (outer diameter, 2 mm; inner diameter, 1.8 mm; length, 25 mm) was inserted. The cannula was secured with 00 silk suture tied snugly around the tracheal circumference.

### *2. Mechanical Properties of the Lung*

A 42-cm long intrapleural catheter was inserted into the sixth or seventh rib interspace in order to measure intrapleural pressure by means of a Validyne differential pressure transducer. The catheter was of polyethylene tubing, 1.8 mm I.D. The rat was placed in a cylindrical body plethysmograph made of plexiglass, and a mesh screen pneumotachometer was installed on one end of the chamber. Pressure differences corresponding to air flow across the screen were measured by a Statham pressure transducer and were then integrated into tidal volume. The tracheal cannula was attached via a hole to the outside and was continuously flushed by either room air or gas mixture at a rate of 1.5 l/min. The respiratory movement of the rat caused a continuous flow of air in and out of the plethysmograph via the pneumotachometer; thus the air inside the chamber was rapidly equilibrated with ambient temperature. Pulmonary compliance was computed (tidal vol/transpulmonary pressure) at zero flow between inspiration and expiration. Pulmonary resistance was calculated by the following equation: $(P1 - P2) \div (F1 - F2)$. P1 and P2 are intrapleural pressures at isovolumetric points on either side of the respiration cycle and F1 and F2 are flows at isovolumetric points. These values were recorded by a Pulmonary Mechanics Computer (Buxco-Electronics).

### *3. Arterial Pressure*

One carotid artery was exposed and a polyethylene cannula (outer diameter, 0.096 cm, inner diameter, 0.058 cm, length 25 cm) was inserted. The cannula was filled with 1.0% heparinized saline and was attached at the opposite end to a Statham pressure transducer. Arterial pressure was recorded by the pulmonary mechanics computer.

### *4. Inhalation of Gas Mixture*

Isopropanol-air mixtures were administered in the following doses: 3.9, 19.5, and 97.5 mg/l. Exposure to the gas mixtures was either nasal or tracheal, according to the group of rats tested. Tracheal exposures were performed by continuously flushing the opening of the tracheal cannula with the appropriate gas mixture at a rate of 1.5 l/min. Rats exposed to the gas mixtures nasally did not require tracheal cannulation; instead, their heads protruded from the body plethysmograph through a rubber diaphragm at one end of the chamber. A plastic dome surrounding the head of each of these rats was flushed with the appropriate gas mixture again at a rate of 1.5 l/min. The gas-air mixtures were administered to all rats for 3 min, alternately with room air for 10 min.

*5. Additional Procedures*

Techniques for measurements of liver and cardiac function performed in some rats are described under Results. All the results were analyzed statistically by the Student's t-test. A P value <0.05 was regarded as significant.

**B. Results**

*1. Effects of Tracheal Inhalation on Pulmonary Mechanics*

Three-minute exposures to three different isopropanol-air mixtures were conducted with three groups of five rats. The following four pulmonary mechanical parameters were recorded: pulmonary resistance, pulmonary compliance, respiratory rate, and tidal volume. For each trial and for each of the four parameters, values were recorded corresponding to the initiation and termination of the exposure periods. In each instance, percentage change has been computed. For each group of five animals, mean percentage change was determined for each parameter and standard error and P values were calculated (Tables 3.1 to 3.3)

In the first group of rats, three concentrations of isopropanol vapor in air were employed: 3.9 mg/l, 19.5 mg/l, and 97.5 mg/l. It is clear that the mixture with the largest concentration of isopropanol produced the most pronounced changes in pulmonary function, while the two less concentrated mixtures exerted little, if any, action on the parameters studied. A mixture of 97.5 mg of isopropanol per liter of air produced an increase in pulmonary resistance and a fall in pulmonary compliance, respiratory rate, and tidal volume. Only the increase in resistance was statistically significant (Table 3.1).

In the second group of five rats, the interaction between isopropanol and a sympathomimetic drug, isoproterenol, was examined. Isoproterenol in a dose of 1 $\mu$g/kg was injected into the femoral vein, while the rat was inhaling room air and also during the inhalation of isopropanol-air mixtures.

As demonstrated by the control animals, isoproterenol evoked a reduction in pulmonary resistance and substantial increases in pulmonary compliance, respiratory rate, and tidal volume. Pulmonary mechanical changes in response to isoproterenol during exposure to 3.9 mg/l isopropanol in air did not differ significantly from those observed in the control animals. During exposure to 19.5 mg/l isopropanol, changes in pulmonary resistance, pulmonary compliance, and tidal volume differed slightly from those of the controls, while the acceleration of respiratory rate in response to the beta-adrenergic stimulant was notably enhanced. Finally, exposure to the most concentrated of the isopropanol-in-air mixtures (97.5 mg/l) potentiated the positive responses of respiratory rate and tidal volume to isoproterenol, though the changes in resistance and compliance differed only moderately from the control responses (Table 3.2).

The third group of five rats was used to examine the role of cholinergic receptors in the airways. In this experiment, only the most concentrated of the isopropanol-air mixtures (97.5 mg/l) was administered though the technical aspects of these trials differed in no way from those before them. After first obtaining a control response to the gas mixture, the experiment was repeated in animals previously injected with atropine (1 mg/kg), a cholinergic blocker. The atropine solution was administered in the right femoral vein. Pretreatment with atropine dramatically reduced the response, in all parameters measured, to the inhaled isopropanol. (Table 3.3).

*2. Effects of Tracheal and Nasal Inhalation on Respiration and Blood Pressure*

Since the experiments described above consisted of administering isopropanol vapor directly to the trachea bypassing the upper respiratory tract, the next group of experiments included rats without tracheal cannulation. It was, therefore, possible to compare one group of rats subjected to nasal inhalation and a second group to tracheal inhalation of isopropanol. In these animals, arterial pressure and the following three pulmonary mechanical parameters were determined: tidal volume, respiratory rate, and minute volume. Tidal volume and respiratory rate were determined as described previously. Minute volume was computed by the pulmonary mechanics computer (minute volume = tidal volume × respiratory rate). For the determination of arterial pressure, the left common carotid artery was pierced and cannulated with

TABLE 3.1

**Respiratory Effects of Isopropanol: Tracheal Inhalation by Rats**

| Procedure | Rat no. | Body wt. (g) | Pulmonary resistance ($cm H_2O/ml/sec$) | | | Pulmonary compliance ($ml/cm H_2O$) | | | Respiratory rate (/min) | | | Tidal volume (ml) | | |
|---|---|---|---|---|---|---|---|---|---|---|---|---|---|---|
| | | | Before | During | %Δ | Before | During | %Δ | Before | During | %Δ | Before | During | %Δ |
| Isopropanol | 1 | 230 | 0.286 | 0.279 | −2.3 | 0.144 | 0.138 | −4.3 | 122.5 | 102.5 | −16.3 | 1.360 | 1.280 | −5.9 |
| 3 min | 2 | 242 | 0.333 | 0.359 | +8.0 | 0.225 | 0.225 | 0.0 | 55.0 | 57.5 | +4.5 | 1.216 | 1.280 | +5.3 |
| 3.9 mg/l | 3 | 210 | 0.333 | 0.354 | +6.4 | 0.178 | 0.172 | −3.5 | 72.5 | 63.5 | −12.4 | 1.040 | 0.960 | −7.6 |
| | 4 | 220 | 0.261 | 0.253 | −3.1 | 0.160 | 0.169 | +5.5 | 68.0 | 65.0 | −4.4 | 1.280 | 1.320 | +3.1 |
| | 5 | 226 | 0.352 | 0.333 | −5.7 | 0.172 | 0.156 | −9.1 | 87.5 | 87.5 | 0.0 | 1.064 | 1.064 | 0.0 |
| | Mean | 225.6 | 0.313 | 0.316 | +1.0 | 0.176 | 0.172 | −4.5 | 81.1 | 75.2 | −5.7 | 1.192 | 1.181 | −1.0 |
| | S.E. | 5.3 | 0.016 | 0.020 | 2.7 | 0.012 | 0.013 | 1.5 | 11.6 | 8.5 | 3.8 | 0.061 | 0.071 | 2.3 |
| | P | | | | NS | | | NS | | | NS | | | NS |
| Isopropanol | 1 | 230 | 0.286 | 0.286 | 0 | 0.141 | 0.134 | −4.4 | 126.25 | 102.5 | −18.8 | 1.240 | 1.140 | −8.1 |
| 3 min | 2 | 242 | 0.359 | 0.375 | +4.4 | 0.225 | 0.225 | 0.0 | 50.0 | 47.5 | −5.0 | 1.328 | 1.360 | +2.4 |
| 19.5 mg/l | 3 | 210 | 0.346 | 0.351 | +1.5 | 0.188 | 0.188 | 0.0 | 61.5 | 63.0 | +2.4 | 0.920 | 0.960 | −4.3 |
| | 4 | 220 | 0.239 | 0.239 | 0.0 | 0.169 | 0.164 | −3.0 | 63.0 | 68.0 | +4.8 | 1.320 | 1.344 | +1.8 |
| | 5 | 226 | 0.326 | 0.313 | −4.1 | 0.163 | 0.163 | 0.0 | 80.8 | 87.5 | +9.4 | 1.224 | 1.064 | −13.1 |
| | Mean | 225.6 | 0.311 | 0.313 | +0.4 | 0.177 | 0.175 | −1.5 | 76.1 | 73.7 | −1.4 | 1.206 | 1.174 | −4.3 |
| | S.E. | 5.3 | 0.021 | 0.023 | 1.4 | 0.013 | 0.014 | 0.9 | 13.4 | 9.6 | 4.9 | 0.074 | 0.078 | 2.9 |
| | P | | | | NS | | | NS | | | NS | | | NS |
| Isopropanol | 1 | 230 | 0.326 | 0.319 | −2.0 | 0.125 | 0.113 | −10.0 | 125.0 | 119.0 | −4.8 | 1.060 | 1.080 | +1.9 |
| 3 min | 2 | 242 | 0.390 | 0.434 | +11.3 | 0.238 | 0.230 | −3.1 | 43.0 | 38.3 | −11.0 | 1.600 | 1.600 | 0.0 |
| 97.5 mg/l | 3 | 210 | 0.394 | 0.505 | +28.4 | 0.194 | 0.203 | +4.5 | 60.0 | 61.5 | +2.5 | 1.120 | 0.880 | −21.4 |
| | 4 | 220 | 0.246 | 0.312 | +27.0 | 0.237 | 0.150 | −36.8 | 72.5 | 60.0 | −17.2 | 1.408 | 1.120 | −20.4 |
| | 5 | 226 | 0.342 | 0.399 | +16.5 | 0.300 | 0.288 | −4.2 | 86.25 | 73.75 | −14.5 | 1.120 | 0.960 | −14.3 |
| | Mean | 225.6 | 0.399 | 0.412 | +16.2 | 0.219 | 0.197 | −10.0 | 77.3 | 70.5 | −9.0 | 1.262 | 1.128 | −10.8 |
| | S.E. | 5.3 | 0.026 | 0.040 | 5.5 | 0.029 | 0.030 | 7.06 | 13.9 | 13.4 | 3.5 | 0.103 | 0.125 | 4.9 |
| | P | | | | 0.05 | | | NS | | | NS | | | NS |

TABLE 3.2

**Interaction of Isopropanol and a Sympathomimetic Drug: Tracheal Inhalation in Rats**

| Procedure | Rat no. | Body wt. (g) | Pulmonary resistance ($cm H_2O/ml/sec$) Before | During | %Δ | Pulmonary compliance ($ml/cm H_2O$) Before | During | %Δ | Respiratory rate (/min) Before | During | %Δ | Tidal volume (ml) Before | During | %Δ |
|---|---|---|---|---|---|---|---|---|---|---|---|---|---|---|
| *Control* – | 1 | 230 | 0.322 | 0.293 | −11.3 | 0.119 | 0.150 | +26.3 | 120.0 | 165.0 | +37.5 | 0.960 | 1.040 | +8.3 |
| Isoproterenol | 2 | 242 | 0.313 | 0.279 | −10.6 | 0.213 | 0.250 | +17.6 | 48.8 | 62.5 | +26.1 | 1.200 | 1.440 | +20.0 |
| 1 μg/kg | 3 | 210 | 0.367 | 0.306 | −16.7 | 0.144 | 0.219 | +52.2 | 65.0 | 81.0 | +24.6 | 1.016 | 1.320 | +29.9 |
| | 4 | 220 | 0.257 | 0.239 | −6.7 | 0.169 | 0.181 | +7.4 | 60.0 | 70.0 | +16.7 | 1.320 | 1.480 | +12.1 |
| | 5 | 245 | 0.293 | 0.239 | −18.2 | 0.141 | 0.200 | +41.6 | 113.0 | 140.0 | +23.9 | 0.960 | 1.064 | +10.8 |
| | Mean | 229.4 | 0.310 | 0.271 | −12.7 | 0.157 | 0.200 | +29.0 | 81.4 | 103.7 | +25.8 | 1.091 | 1.269 | +16.2 |
| | S.E. | 6.6 | 0.017 | 0.013 | 2.1 | 0.015 | 0.017 | 8.1 | 14.6 | 20.5 | 3.3 | 0.072 | 0.092 | 3.9 |
| | P | | | | 0.01 | | | 0.02 | | | 0.02 | | | 0.01 |
| Isopropanol | 1 | 230 | 0.279 | 0.266 | −4.8 | 0.138 | 0.156 | +13.6 | 102.5 | 140.0 | +36.6 | 1.280 | 1.400 | +9.4 |
| 3.9 mg/l | 2 | 242 | 0.359 | 0.299 | −16.7 | 0.225 | 0.255 | +13.3 | 57.5 | 65.0 | +13.0 | 1.280 | 1.472 | +15.0 |
| 3 min | 3 | 210 | 0.354 | 0.274 | −22.5 | 0.172 | 0.244 | +41.8 | 63.5 | 92.5 | 45.7 | 0.960 | 1.240 | +29.1 |
| Isoproterenol | 4 | 220 | 0.253 | 0.230 | −8.9 | 0.169 | 0.200 | +18.5 | 65.0 | 77.5 | +16.1 | 1.320 | 1.640 | +24.2 |
| 1 μg/kg | 5 | 245 | 0.243 | 0.226 | −7.1 | 0.170 | 0.212 | +25.0 | 152.5 | 190.0 | +24.6 | 1.040 | 1.160 | +11.5 |
| | Mean | 229.4 | 0.298 | 0.259 | −12.0 | 0.175 | 0.213 | +22.4 | 88.2 | 113.0 | +27.2 | 1.176 | 1.382 | +17.8 |
| | S.E. | 6.6 | 0.024 | 0.013 | 3.3 | 0.013 | 0.017 | 5.3 | 17.9 | 23.1 | 6.1 | 0.073 | 0.085 | 3.8 |
| | P | | | | 0.05 | | | 0.01 | | | 0.02 | | | 0.01 |
| Isopropanol | 1 | 230 | 0.286 | 0.269 | −5.8 | 0.134 | 0.150 | +11.6 | 102.5 | 167.5 | +63.4 | 1.140 | 1.140 | 0.0 |
| 19.5 mg/l | 2 | 242 | 0.375 | 0.314 | −16.3 | 0.225 | 0.250 | +11.1 | 47.5 | 62.5 | +31.6 | 1.360 | 1.440 | +58.8 |
| 3 min | 3 | 210 | 0.351 | 0.299 | −14.8 | 0.188 | 0.240 | +28.0 | 63.0 | 96.5 | +53.1 | 0.960 | 1.160 | +20.8 |
| Isoproterenol | 4 | 220 | 0.239 | 0.213 | −11.1 | 0.164 | 0.206 | +25.9 | 66.0 | 82.5 | +25.0 | 1.344 | 1.776 | +32.7 |
| 1 μg/kg | 5 | 245 | 0.226 | 0.206 | −8.8 | 0.194 | 0.219 | +12.9 | 150.0 | 210.0 | +40.0 | 1.120 | 1.200 | +7.1 |
| | Mean | 229.4 | 0.295 | 0.260 | −11.4 | 0.181 | 0.213 | +17.9 | 85.8 | 123.8 | +42.6 | 1.185 | 1.343 | +23.9 |
| | S.E. | 6.6 | 0.029 | 0.021 | 1.9 | 0.015 | 0.017 | 3.7 | 18.4 | 27.9 | 7.0 | 0.075 | 0.120 | 10.4 |
| | P | | | | 0.02 | | | 0.01 | | | 0.02 | | | NS |
| Isopropanol | 1 | 230 | 0.319 | 0.259 | −18.7 | 0.113 | 0.154 | +36.7 | 109.0 | 132.5 | +21.3 | 1.080 | 1.200 | +11.1 |
| 97.5 mg/l | 2 | 242 | 0.434 | 0.386 | −11.0 | 0.230 | 0.265 | +15.2 | 38.3 | 53.3 | +39.2 | 1.536 | 1.600 | +4.2 |
| 3 min | 3 | 210 | 0.505 | 0.379 | −25.0 | 0.203 | 0.244 | +20.4 | 61.5 | 107.5 | +74.8 | 0.880 | 1.240 | +40.9 |
| Isoproterenol, | 4 | 220 | 0.312 | 0.246 | −21.3 | 0.150 | 0.210 | +40.0 | 60.0 | 80.0 | +33.3 | 1.120 | 1.880 | +67.8 |
| 1 μg/kg | 5 | 245 | 0.266 | 0.246 | −7.5 | 0.150 | 0.162 | +8.3 | 150.0 | 197.5 | +31.7 | 0.880 | 1.040 | +18.2 |
| | Mean | 229.4 | 0.367 | 0.303 | −16.7 | 0.169 | 0.207 | +24.12 | 85.7 | 114.2 | +40.1 | 1.112 | 1.328 | +28.4 |
| | S.E. | 6.6 | 0.044 | 0.032 | 3.2 | 0.020 | 0.021 | 6.1 | 20.9 | 24.7 | 9.15 | 0.132 | 0.144 | 11.6 |
| | P | | | | 0.02 | | | 0.01 | | | 0.01 | | | NS |

TABLE 3.3

**Respiratory Effects of Isopropanol During Cholinergic Blockade: Tracheal Inhalation in Rats**

| | | | Pulmonary resistance (cm $H_2O$/ml/sec) | | | Pulmonary compliance (ml/cm $H_2O$) | | | Respiratory rate (/min) | | | Tidal volume (ml) | | |
|---|---|---|---|---|---|---|---|---|---|---|---|---|---|---|
| Procedure | Rat no. | Body wt. (g) | Before | During | %Δ | Before | During | %Δ | Before | During | %Δ | Before | During | %Δ |
| Isopropanol | 1 | 230 | 0.326 | 0.319 | −2.0 | 0.125 | 0.113 | −10.0 | 125.0 | 119.0 | −4.8 | 1.060 | 1.080 | +1.9 |
| 97.5 mg/l | 2 | 242 | 0.390 | 0.434 | +11.3 | 0.238 | 0.230 | −3.1 | 43.0 | 38.3 | −11.0 | 1.600 | 1.600 | 0.0 |
| 3 min | 3 | 210 | 0.394 | 0.505 | +28.4 | 0.194 | 0.203 | +4.5 | 60.0 | 61.5 | +2.5 | 1.120 | 0.880 | −21.4 |
| | 4 | 220 | 0.246 | 0.312 | +27.0 | 0.237 | 0.150 | −36.8 | 72.5 | 60.0 | −17.2 | 1.408 | 1.120 | −20.4 |
| | 5 | 226 | 0.342 | 0.399 | +16.5 | 0.300 | 0.288 | −4.2 | 86.25 | 73.75 | −14.5 | 1.120 | 0.960 | −14.3 |
| | Mean | 225.6 | 0.339 | 0.412 | +16.2 | 0.219 | 0.197 | −10.0 | 77.3 | 70.5 | −9.0 | 1.262 | 1.128 | −10.8 |
| | S.E. | 5.3 | 0.026 | 0.040 | 5.5 | 0.029 | 0.030 | 7.06 | 13.9 | 13.4 | 3.5 | 0.103 | 0.125 | 4.9 |
| | P | | | | 0.05 | | | NS | | | NS | | | 0.05 |
| Atropine | 1 | 220 | 0.219 | 0.226 | +3.0 | 0.187 | 0.181 | −3.3 | 68.0 | 66.5 | −2.2 | 1.440 | 1.400 | −2.8 |
| 1 mg/kg | 2 | 224 | 0.323 | 0.332 | +2.9 | 0.112 | 0.110 | −2.2 | 32.5 | 33.0 | +1.5 | 1.360 | 1.360 | 0.0 |
| Isopropanol | 3 | 218 | 0.286 | 0.293 | +2.3 | 0.156 | 0.154 | −1.6 | 42.5 | 45.0 | +8.2 | 1.760 | 1.816 | +3.2 |
| 97.5 mg/l | 4 | 220 | 0.247 | 0.266 | +7.5 | 0.200 | 0.187 | −6.2 | 50.0 | 52.5 | +5.0 | 1.360 | 1.280 | −6.2 |
| 3 min | 5 | 230 | 0.319 | 0.319 | 0.0 | 0.153 | 0.132 | −13.4 | 75.0 | 71.5 | −4.7 | 2.352 | 2.272 | −3.4 |
| | Mean | 222.4 | 0.279 | 0.287 | +3.14 | 0.162 | 0.153 | −5.34 | 53.60 | 53.70 | +1.56 | 1.654 | 1.626 | −1.84 |
| | S.E. | 2.1 | 0.019 | 0.018 | 1.22 | 0.015 | 0.014 | 2.16 | 7.90 | 7.02 | 2.34 | 0.189 | 0.186 | 1.59 |
| | P | | | | NS | | | NS | | | NS | | | NS |

polyethylene tubing. The tubing was filled with heparinized (1%) saline and attached to a Statham pressure transducer. The transducer signal was amplified and recorded by the pulmonary mechanics computer.

The results of tracheal inhalation are summarized in Table 3.4. It is noted that the two lesser concentrations of the isopropanol-air mixtures applied produced no significant adjustments in the parameters observed. Tidal volume, respiratory rate, and arterial pressure were depressed after 3-min exposures to the most concentrated mixture (94.5 mg/l) but only the fall in blood pressure was significant.

Isopropanol-air mixtures were inhaled nasally and the results are summarized in Table 3.5. Here, the intermediate (19.5 mg/l) of the three concentrations administered caused a depression of pulmonary function, most notably of minute volume and tidal volume. The results show a trend of reduction in minute volume and arterial pressure. The depression of all parameters observed was substantially enhanced when the most concentrated of the mixtures (97.5 mg/l) was applied. Tidal volume, respiratory rate, minute volume, and arterial pressure were depressed 20.3%, 13.5%, 30.9%, and 28.6%, respectively. That the effects of nasal inhalation are more intense than those of tracheal inhalation indicates that the passage of isopropanol to the upper respiratory tract was contributing to the depression of respiration and to the hypotension.

The final step in the experiments involving nasal and tracheal inhalation was to preadminister atropine (1 mg/kg) intravenously. The responses to tracheal inhalation and nasal inhalation of isopropanol noted earlier were eliminated in the two groups of animals that were pretreated with atropine (Tables 3.6 and 3.7).

*3. Effects of Tracheal and Nasal Inhalation on the Electrocardiogram*

The Lead II electrocardiogram was recorded from rats during inhalational exposure to isopropanol, simultaneous to the recording of pulmonary mechanical functions and blood pressure. Electrocardiograms were obtained from rats involved in experiments described in Tables 3.4 to 3.7. Heart rate (beats/min) and PR interval (msec) were measured from the electrocardiogram at points corresponding to elapsed times of 3 min and 5 min after exposure.

Heart rate was measured in animals exposed to three concentrations of isopropanol-in-air (3.9, 19.5, and 97.5 mg/l) by either nasal or tracheal routes. These trials were repeated with rats pretreated with atropine (1 mg/kg, intravenously). Significant changes in rate were not apparent in rats treated with the two lower concentrations of the alcohol. Animals exposed tracheally to the most concentrated of the gas mixtures displayed a moderate depression of heart rate, which was diminished when the rats were pretreated with atropine (Table 3.8). PR intervals were measured on electrocardiograms (Table 3.9). No significant pattern of response is discernible. Two trials produced statistically significant results, though these corresponded to rather small relative changes in the PR interval.

*4. Effects of Subacute Inhalation on Responsiveness of the Airways*

Rats were exposed to a low concentration of isopropanol-in-air (400 ppm or 0.098 mg/l) for 1 hr each day for 6 days, and were then subjected to acute exposures to isopropanol-air mixtures, as described earlier, on the seventh day. Daily exposures were divided into two periods of 1/2 hr. The animals were placed in groups of four into glass bell jars, through which was passed a 400 ppm isopropanol-air mixture at a rate of 1.5 l/min. They were prepared for study on the seventh day and the following four pulmonary mechanical parameters were measured: pulmonary resistance, pulmonary compliance, respiratory rate, and tidal volume. Each animal was exposed for 3 min to gas mixtures of 3.9, 19.5, and 97.5 mg isopropanol per liter of air.

When these results are compared to those found in Table 3.1, it is obvious that 6 days of subacute exposure to isopropanol vapor produced profound changes in the acute responses of these animals to higher concentrations of the solvent. The animals tested with no previous exposure showed pulmonary resistance changes of +0.4% and +16.2% when challenged with isopropanol-air mixtures of concentrations 19.5 and 97.5 mg/l, respectively. This compares with changes of +7.4% and +31.7%, on exposure to the same gas concentrations, for the rats that had been exposed to the solvent for 6 previous days. Thus, the subacute exposures induced a hyperresponsiveness of airway constriction in the acute response to the inhaled solvent. Similarly, pulmonary compliance changes in

TABLE 3.4

**Respiratory and Circulatory Effects of Isopropanol: Tracheal Inhalation in Rats**

| Procedure | Rat no. | Body wt. (g) | Tidal volume (ml) Before | Tidal volume (ml) 3 min | Tidal volume (ml) %Δ | Respiratory rate (/min) Before | Respiratory rate (/min) 3 min | Respiratory rate (/min) %Δ | Minute volume (ml) Before | Minute volume (ml) 3 min | Minute volume (ml) %Δ | Arterial pressure (mmHg) Before | Arterial pressure (mmHg) 3 min | Arterial pressure (mmHg) %Δ |
|---|---|---|---|---|---|---|---|---|---|---|---|---|---|---|
| Isopropanol (3.9 mg/l) | 1 | 310 | 1.875 | 1.775 | −5.3 | 91.5 | 90.0 | −1.6 | 171.6 | 159.7 | −6.9 | 140.0 | 139.0 | −0.7 |
| | 2 | 280 | 2.212 | 2.262 | +2.2 | 76.0 | 75.3 | −0.9 | 168.1 | 170.3 | +1.3 | 113.2 | 110.0 | −2.9 |
| | 3 | 245 | 1.265 | 1.290 | +2.0 | 104.0 | 105.0 | +1.0 | 131.6 | 135.4 | +2.9 | 90.0 | 90.0 | 0.0 |
| | 4 | 250 | 1.350 | 1.350 | 0.0 | 52.5 | 50.0 | −4.8 | 70.9 | 67.5 | −4.8 | 126.0 | 130.0 | +3.2 |
| | 5 | 265 | 1.787 | 1.775 | −0.7 | 66.0 | 66.5 | +0.7 | 117.9 | 118.0 | +0.1 | 157.5 | 160.0 | +1.6 |
| | Mean | 270.0 | 1.698 | 1.690 | −0.4 | 78.0 | 77.4 | −1.1 | 132.0 | 130.2 | −1.5 | 125.3 | 125.8 | +0.2 |
| | S.E. | 11.7 | 0.175 | 0.175 | 1.3 | 9.1 | 9.4 | 1.0 | 18.4 | 18.1 | 1.9 | 11.5 | 12.0 | 1.0 |
| | P | | | | NS | | | NS | | | NS | | | NS |
| Isopropanol (19.5 mg/l) | 1 | 280 | 2.312 | 2.375 | +2.7 | 75.0 | 75.3 | +0.4 | 173.4 | 178.8 | +3.1 | 110.0 | 111.5 | +1.4 |
| | 2 | 250 | 1.400 | 1.475 | +5.3 | 57.5 | 49.5 | −13.9 | 80.5 | 73.0 | −9.3 | 130.0 | 121.0 | −6.9 |
| | 3 | 245 | 1.150 | 1.080 | −6.1 | 45.2 | 48.8 | +8.0 | 52.0 | 52.7 | +1.3 | 115.0 | 111.5 | −3.0 |
| | 4 | 265 | 1.850 | 1.750 | −5.4 | 68.0 | 68.5 | +0.7 | 125.8 | 119.9 | −4.7 | 155.0 | 161.5 | +4.2 |
| | 5 | 310 | 1.687 | 1.587 | −5.9 | 100.0 | 99.0 | −1.0 | 168.7 | 157.1 | −6.9 | 140.0 | 135.0 | −3.6 |
| | Mean | 270.0 | 1.680 | 1.653 | −1.9 | 69.1 | 68.2 | −1.2 | 120.1 | 116.3 | −3.3 | 130.0 | 128.1 | −1.6 |
| | S.E. | 11.7 | 0.198 | 0.211 | 2.4 | 9.2 | 9.3 | 3.5 | 23.9 | 24.0 | 2.4 | 8.2 | 9.4 | 1.9 |
| | P | | | | NS | | | NS | | | NS | | | NS |
| Isopropanol (97.5 mg/l) | 1 | 310 | 0.900 | 0.775 | −16.1 | 92.5 | 81.4 | −12.0 | 83.25 | 63.1 | −24.2 | 111.0 | 95.0 | −14.4 |
| | 2 | 280 | 2.350 | 2.350 | 0.0 | 76.0 | 80.5 | +5.9 | 178.6 | 189.2 | +5.9 | 85.0 | 70.4 | −17.2 |
| | 3 | 245 | 1.020 | 1.000 | −2.0 | 52.6 | 53.0 | +0.8 | 53.6 | 53.0 | −1.1 | 107.5 | 92.5 | −13.9 |
| | 4 | 265 | 1.800 | 1.750 | −2.8 | 69.0 | 70.0 | +1.4 | 124.2 | 122.5 | −1.4 | 181.0 | 154.0 | −14.9 |
| | 5 | 250 | 1.415 | 1.200 | −15.2 | 57.5 | 58.0 | +0.9 | 81.4 | 69.6 | −14.5 | 133.5 | 128.5 | −3.7 |
| | Mean | 270.0 | 1.497 | 1.415 | −7.2 | 69.5 | 68.6 | −0.6 | 104.2 | 99.5 | −7.1 | 123.4 | 108.1 | −12.8 |
| | S.E. | 11.7 | 0.265 | 0.284 | 3.5 | 7.1 | 5.7 | 3.0 | 21.7 | 25.4 | 5.4 | 16.3 | 14.8 | 2.3 |
| | P | | | | NS | | | NS | | | NS | | | 0.01 |

TABLE 3.5

**Respiratory and Circulatory Effects of Isopropanol: Nasal Inhalation in Rats**

| Procedure | Rat no. | Body wt. (g) | Tidal volume (ml) Before | Tidal volume (ml) 3 min | Tidal volume (ml) %Δ | Respiratory rate (/min) Before | Respiratory rate (/min) 3 min | Respiratory rate (/min) %Δ | Minute volume (ml) Before | Minute volume (ml) 3 min | Minute volume (ml) %Δ | Arterial pressure (mmHg) Before | Arterial pressure (mmHg) 3 min | Arterial pressure (mmHg) %Δ |
|---|---|---|---|---|---|---|---|---|---|---|---|---|---|---|
| Nasal exposure (3.9 mg/l) | 1 | 310 | 1.375 | 1.437 | +4.5 | 106.5 | 105.0 | −1.4 | 146.4 | 150.9 | +3.1 | 123.5 | 120.0 | −2.8 |
| | 2 | 275 | 1.300 | 1.412 | +8.6 | 73.6 | 74.0 | +0.5 | 95.7 | 104.5 | +9.2 | 139.0 | 137.5 | −1.1 |
| | 3 | 290 | 1.825 | 1.812 | −0.7 | 65.0 | 61.2 | −5.8 | 118.6 | 110.9 | −6.5 | 113.5 | 116.5 | +2.6 |
| | 4 | 250 | 1.350 | 1.250 | −7.4 | 64.0 | 62.0 | −3.1 | 86.4 | 77.5 | −10.3 | 132.5 | 135.0 | +1.9 |
| | 5 | 265 | 1.562 | 1.375 | −12.0 | 98.5 | 98.0 | −0.5 | 153.8 | 134.7 | −12.4 | 142.5 | 142.5 | 0.0 |
| | Mean | 278.0 | 1.482 | 1.457 | −1.4 | 81.5 | 80.0 | −2.1 | 120.2 | 115.7 | −3.4 | 130.2 | 130.3 | +0.12 |
| | S.E. | 10.3 | 0.096 | 0.094 | 3.8 | 8.8 | 9.1 | 1.1 | 13.3 | 12.7 | 4.1 | 5.3 | 5.1 | 1.0 |
| | P | | | | NS | | | NS | | | NS | | | NS |
| Nasal exposure (19.5 mg/l) | 1 | 310 | 1.437 | 1.375 | −4.8 | 110.0 | 95.0 | −13.6 | 158.1 | 130.6 | −17.4 | 118.5 | 105.0 | −11.4 |
| | 2 | 275 | 1.700 | 1.585 | −6.8 | 78.0 | 78.0 | 0 | 132.6 | 123.6 | −6.8 | 150.0 | 140.0 | −6.7 |
| | 3 | 290 | 1.875 | 1.837 | −2.0 | 61.0 | 62.6 | +2.6 | 114.4 | 115.0 | +0.5 | 116.5 | 115.0 | −1.3 |
| | 4 | 250 | 1.315 | 1.160 | −11.8 | 62.6 | 59.0 | −5.7 | 82.3 | 68.4 | −16.9 | 152.5 | 132.0 | −13.4 |
| | 5 | 265 | 1.687 | 1.500 | −11.1 | 102.5 | 90.0 | −12.2 | 172.9 | 135.0 | −21.9 | 138.0 | 140.0 | +1.4 |
| | Mean | 278.0 | 1.603 | 1.491 | −7.3 | 82.8 | 76.9 | −5.8 | 132.1 | 114.5 | −12.5 | 135.1 | 126.4 | −6.3 |
| | S.E. | 10.3 | 0.100 | 0.112 | 1.9 | 10.1 | 7.2 | 3.2 | 16.0 | 12.0 | 4.1 | 7.6 | 7.0 | 2.8 |
| | P | | | | 0.02 | | | NS | | | NS | | | NS |
| Nasal exposure (97.5 mg/l) | 1 | 310 | 1.500 | 1.300 | −13.3 | 100.0 | 80.0 | −20.0 | 150.0 | 104.0 | −30.7 | 115.0 | 85.0 | −26.1 |
| | 2 | 275 | 1.262 | 1.037 | −17.8 | 80.0 | 78.0 | −2.5 | 101.0 | 80.9 | −19.9 | 116.0 | 75.0 | −35.3 |
| | 3 | 290 | 1.875 | 1.567 | −16.7 | 59.0 | 55.0 | −6.8 | 110.6 | 85.9 | −22.3 | 110.0 | 83.5 | −24.8 |
| | 4 | 250 | 2.100 | 1.625 | −22.6 | 79.0 | 63.0 | −20.2 | 114.5 | 63.0 | −45.0 | 140.0 | 87.5 | −37.5 |
| | 5 | 265 | 1.450 | 1.000 | −31.0 | 123.0 | 101.0 | −17.9 | 258.3 | 164.1 | −36.5 | 118.0 | 95.0 | −19.5 |
| | Mean | 278.0 | 1.637 | 1.305 | −20.3 | 88.2 | 75.4 | −13.5 | 146.9 | 99.6 | −30.9 | 119.8 | 85.2 | −28.6 |
| | S.E. | 10.3 | 0.152 | 0.129 | 3.1 | 10.8 | 7.9 | 3.7 | 29.1 | 17.4 | 4.6 | 5.2 | 3.2 | 3.4 |
| | P | | | | 0.01 | | | 0.02 | | | 0.02 | | | 0.01 |

TABLE 3.6

Respiratory and Circulatory Effects of Isopropanol During Cholinergic Blockade: Tracheal Inhalation in Rats

| Procedure | Rat no. | Body wt. (g) | Tidal volume (ml) | | | Respiratory rate (/min) | | | Minute volume (ml) | | | Arterial pressure (mmHg) | | |
|---|---|---|---|---|---|---|---|---|---|---|---|---|---|---|
| | | | Before | 3 min | %Δ | Before | 3 min | %Δ | Before | 3 min | %Δ | Before | 3 min | %Δ |
| Isopropanol | 1 | 280 | 2.225 | 2.087 | −6.2 | 77.5 | 74.0 | −4.5 | 172.4 | 154.4 | −10.4 | 121.5 | 123.0 | +1.2 |
| (3.9 mg/l) | 2 | 270 | 2.225 | 2.325 | +2.2 | 61.6 | 62.6 | +1.6 | 137.1 | 145.4 | +6.0 | 157.5 | 160.0 | +1.6 |
| after atropine | 3 | 260 | 0.925 | 0.965 | +4.3 | 88.0 | 92.5 | +5.1 | 81.4 | 89.3 | +9.7 | 147.5 | 150.0 | +1.7 |
| (1 mg/kg, i.v.) | 4 | 265 | 2.500 | 2.250 | −10.0 | 51.4 | 48.0 | −6.6 | 128.5 | 108.0 | −15.9 | 125.0 | 129.0 | +3.2 |
| | 5 | 245 | 1.900 | 1.775 | −6.6 | 100.0 | 105.0 | +5.0 | 190.0 | 186.4 | −1.9 | 165.0 | 162.5 | −1.5 |
| | Mean | 264.0 | 1.955 | 1.880 | −3.3 | 75.7 | 76.4 | +0.1 | 141.9 | 136.7 | −2.5 | 143.3 | 144.9 | +1.24 |
| | S.E. | 5.8 | 0.274 | 0.247 | 2.76 | 8.7 | 10.2 | 2.4 | 18.8 | 17.2 | 4.8 | 8.7 | 8.0 | 0.76 |
| | P | | | | NS | | | NS | | | NS | | | NS |
| Isopropanol | 1 | 280 | 2.100 | 2.087 | −0.6 | 77.5 | 81.5 | +5.2 | 162.7 | 170.1 | +4.5 | 123.5 | 128.5 | +4.0 |
| (19.5 mg/l) | 2 | 270 | 2.187 | 2.250 | +2.8 | 68.6 | 70.0 | +2.0 | 150.0 | 157.5 | +5.0 | 160.0 | 162.5 | +1.6 |
| after atropine | 3 | 260 | 1.075 | 1.100 | +2.3 | 110.0 | 111.5 | +1.4 | 118.2 | 122.6 | +3.7 | 117.0 | 110.0 | −6.0 |
| (1 mg/kg, i.v.) | 4 | 265 | 2.225 | 2.287 | +2.8 | 105.0 | 106.5 | +1.4 | 233.6 | 243.6 | +4.3 | 132.5 | 125.0 | −5.7 |
| | 5 | 245 | 1.737 | 1.750 | +0.7 | 81.5 | 77.5 | −4.9 | 141.6 | 135.6 | −4.2 | 95.0 | 101.0 | +6.3 |
| | Mean | 264.0 | 1.865 | 1.895 | +1.6 | 88.5 | 89.4 | +1.02 | 161.2 | 165.9 | +2.7 | 125.6 | 125.4 | +0.04 |
| | S.E. | 5.8 | 0.215 | 0.220 | 0.67 | 8.06 | 8.2 | 1.64 | 19.5 | 21.1 | 1.7 | 10.6 | 10.5 | 2.5 |
| | P | | | | NS | | | NS | | | NS | | | NS |
| Isopropanol | 1 | 280 | 2.325 | 2.187 | −5.9 | 87.5 | 83.5 | −4.6 | 203.4 | 182.6 | −10.2 | 123.0 | 135.0 | +9.7 |
| (97.5 mg/l) | 2 | 270 | 2.125 | 2.625 | +19.0 | 77.5 | 75.0 | −3.2 | 164.7 | 196.8 | +19.4 | 158.5 | 150.0 | −5.4 |
| after atropine | 3 | 260 | 1.250 | 1.100 | −12.0 | 82.5 | 88.6 | +7.3 | 103.1 | 97.3 | −5.6 | 103.5 | 97.5 | −8.2 |
| (1 mg/kg, i.v.) | 4 | 265 | 2.250 | 2.300 | +2.2 | 110.0 | 121.0 | +10.0 | 247.5 | 278.3 | +12.4 | 119.5 | 112.5 | −5.8 |
| | 5 | 245 | 1.875 | 1.775 | −5.3 | 56.4 | 56.0 | −0.7 | 105.7 | 99.4 | −6.0 | 127.5 | 123.0 | −3.5 |
| | Mean | 264.0 | 1.965 | 1.997 | −0.4 | 82.8 | 84.8 | +1.76 | 164.9 | 170.8 | +2.0 | 126.4 | 123.6 | −2.6 |
| | S.E. | 5.8 | 0.194 | 0.262 | 5.3 | 8.6 | 10.6 | 2.9 | 27.9 | 33.8 | 5.8 | 9.0 | 9.0 | 3.2 |
| | P | | | | NS | | | NS | | | NS | | | NS |

TABLE 3.7

**Respiratory and Circulatory Effects of Isopropanol During Cholinergic Blockade: Nasal Inhalation in Rats**

| Procedure | Rat no. | Body wt. (g) | Tidal volume (ml) Before | Tidal volume (ml) 3 min | Tidal volume (ml) %Δ | Respiratory rate (/min) Before | Respiratory rate (/min) 3 min | Respiratory rate (/min) %Δ | Minute volume (ml) Before | Minute volume (ml) 3 min | Minute volume (ml) %Δ | Arterial pressure (mmHg) Before | Arterial pressure (mmHg) 3 min | Arterial pressure (mmHg) %Δ |
|---|---|---|---|---|---|---|---|---|---|---|---|---|---|---|
| Isopropanol | 1 | 270 | 1.031 | 1.000 | −3.0 | 70.8 | 71.6 | +1.1 | 73.0 | 78.8 | +7.9 | 132.5 | 133.7 | +0.9 |
| 3.9 mg/l | 2 | 250 | 0.925 | 0.844 | −8.8 | 91.5 | 95.0 | +3.8 | 84.6 | 80.2 | −5.2 | 141.0 | 143.5 | +1.8 |
| after atropine | 3 | 255 | 1.100 | 1.125 | +2.3 | 90.0 | 96.0 | +6.7 | 99.0 | 108.0 | +9.1 | 185.0 | 178.5 | −3.5 |
| (1 mg/kg, i.v.) | 4 | 285 | 1.260 | 1.294 | +2.7 | 76.0 | 72.8 | −4.2 | 95.8 | 94.2 | −1.7 | 112.2 | 109.0 | −2.8 |
| | 5 | 300 | 1.380 | 1.354 | −1.9 | 68.5 | 70.0 | +2.2 | 94.5 | 94.8 | +0.3 | 120.0 | 149.9 | −4.2 |
| | Mean | 272.0 | 1.139 | 1.123 | −1.7 | 79.4 | 81.1 | +1.9 | 89.4 | 91.2 | +2.1 | 138.1 | 135.9 | −1.6 |
| | S.E. | 9.3 | 0.081 | 0.093 | 2.1 | 4.8 | 5.9 | 1.8 | 4.7 | 5.4 | 2.8 | 12.7 | 12.3 | 1.2 |
| | P | | | | NS | | | NS | | | NS | | | NS |
| Isopropanol | 1 | 270 | 1.000 | 1.106 | +10.6 | 78.0 | 77.0 | −1.3 | 78.0 | 85.2 | +9.2 | 163.7 | 165.0 | +0.8 |
| 19.5 mg/l | 2 | 250 | 0.762 | 0.719 | −5.7 | 91.5 | 100.0 | +9.3 | 69.7 | 71.9 | +3.1 | 145.0 | 145.0 | 0 |
| after atropine | 3 | 255 | 1.125 | 1.106 | −1.7 | 101.5 | 100.0 | −1.5 | 114.2 | 110.6 | −3.1 | 182.0 | 182.0 | 0 |
| (1 mg/kg, i.v.) | 4 | 285 | 1.280 | 1.256 | −1.9 | 73.2 | 71.6 | −2.2 | 93.7 | 89.9 | −4.0 | 118.5 | 112.7 | −4.9 |
| | 5 | 300 | 1.340 | 1.366 | +1.9 | 68.0 | 71.3 | +4.8 | 91.1 | 97.4 | +6.9 | 110.0 | 102.4 | −6.9 |
| | Mean | 272.0 | 1.101 | 1.111 | +0.6 | 82.4 | 84.0 | +1.8 | 89.3 | 91.0 | +2.4 | 143.8 | 141.4 | −2.2 |
| | S.E. | 9.3 | 0.104 | 0.109 | 2.8 | 6.2 | 6.6 | 2.2 | 7.6 | 6.4 | 2.6 | 13.5 | 15.1 | 1.5 |
| | P | | | | NS | | | NS | | | NS | | | NS |
| Isopropanol | 1 | 270 | 0.950 | 0.812 | −14.5 | 76.0 | 74.2 | −2.4 | 72.2 | 60.2 | −16.6 | 166.0 | 140.0 | −15.7 |
| 97.5 mg/l | 2 | 250 | 0.810 | 0.775 | −4.3 | 97.0 | 92.5 | −4.6 | 78.6 | 71.7 | −8.8 | 146.5 | 144.5 | −1.4 |
| after atropine | 3 | 255 | 1.225 | 1.200 | −2.0 | 96.5 | 102.5 | +6.2 | 118.2 | 123.0 | +4.1 | 162.5 | 175.0 | +7.7 |
| (1 mg/kg, i.v.) | 4 | 285 | 1.200 | 1.265 | +5.4 | 72.5 | 65.0 | −10.3 | 87.0 | 82.2 | −5.5 | 116.5 | 114.5 | −1.7 |
| | 5 | 300 | 1.450 | 1.375 | −5.2 | 80.0 | 74.0 | −7.5 | 116.0 | 101.7 | −12.3 | 113.5 | 110.0 | −3.1 |
| | Mean | 272.0 | 1.127 | 1.085 | −4.1 | 84.4 | 81.6 | −3.7 | 94.4 | 87.8 | −7.8 | 141.0 | 136.8 | −2.8 |
| | S.E. | 9.3 | 0.112 | 0.122 | 3.2 | 5.2 | 6.9 | 2.8 | 9.6 | 11.1 | 3.5 | 11.1 | 11.7 | 3.7 |
| | P | | | | NS | | | NS | | | NS | | | NS |

TABLE 3.8

**Electrocardiographic Effects of Isopropanol in Rats: Heart Rate**

| Procedure | Rat no. | Body wt. (g) | 3.9 mg/l Initial | 3 min | %Δ | 5 min | %Δ | 19.5 mg/l Initial | 3 min | %Δ | 5 min | %Δ | 97.5 mg/l Initial | 3 min | %Δ | 5 min | %Δ |
|---|---|---|---|---|---|---|---|---|---|---|---|---|---|---|---|---|---|
| Nasal exposure | 1 | 310 | 490 | 480 | −2.0 | 480 | −2.0 | 480 | 480 | 0 | 480 | 0 | 480 | 490 | +2.1 | 480 | 0 |
| | 2 | 275 | 410 | 420 | +2.4 | 420 | +2.4 | 430 | 390 | −9.3 | 400 | −6.9 | 390 | 400 | +2.6 | 400 | +2.6 |
| | 3 | 290 | 420 | 410 | −2.4 | 410 | −2.4 | 400 | 400 | 0 | 400 | 0 | 400 | 390 | −2.5 | 410 | +2.5 |
| | 4 | 250 | 430 | 440 | +2.3 | 460 | +6.9 | 460 | 460 | 0 | 450 | −2.2 | 420 | 440 | +4.8 | 430 | +2.3 |
| | 5 | 265 | 520 | 530 | +1.9 | 510 | −1.9 | 460 | 470 | +2.2 | 430 | −6.5 | 440 | 470 | +6.8 | 490 | +11.4 |
| | Mean | 278.0 | 454 | 456 | +0.44 | 456 | +0.6 | 437.5 | 440 | −1.4 | 432 | −3.1 | 426 | 438 | +2.8 | 442 | +3.7 |
| | S.E. | 10.3 | 21.6 | 22.0 | 1.1 | 18.6 | 1.8 | 14.3 | 18.7 | 2.0 | 15.3 | 1.5 | 16.0 | 19.3 | 1.6 | 18.3 | 1.9 |
| | P | | | | NS | | NS | | | NS | | NS | | | NS | | NS |
| Nasal exposure after atropine (1 mg/kg, i.v.) | 1 | 270 | 460 | 470 | +2.2 | 480 | +4.2 | 485 | 485 | 0 | 480 | −1.0 | 480 | 460 | −4.2 | 470 | −2.1 |
| | 2 | 250 | 390 | 385 | −1.3 | 390 | 0 | 395 | 400 | +1.3 | 400 | +1.3 | 400 | 400 | 0 | 390 | −2.5 |
| | 3 | 255 | 400 | 400 | 0 | 385 | −3.7 | 390 | 410 | +5.1 | 410 | +5.1 | 430 | 415 | −3.5 | 420 | −2.3 |
| | 4 | 285 | 430 | 420 | −2.3 | 410 | −4.6 | 420 | 410 | −2.4 | 420 | 0 | 420 | 410 | −2.4 | 410 | −2.4 |
| | 5 | 300 | 420 | 420 | 0 | 415 | −1.2 | 420 | 430 | −2.4 | 420 | 0 | 410 | 400 | −2.4 | 400 | −2.4 |
| | Mean | 272.0 | 420 | 419 | −0.3 | 416 | −1.1 | 422 | 427 | +0.3 | 426 | +1.1 | 428 | 417 | −2.5 | 418 | −2.3 |
| | S.E. | 9.3 | 12.2 | 14.3 | 0.7 | 16.9 | 1.5 | 16.9 | 15.3 | 1.4 | 14.0 | 1.1 | 13.9 | 11.1 | 0.7 | 13.9 | 0.1 |
| | P | | | | NS | | NS | | | NS | | NS | | | 0.02 | | 0.001 |
| Tracheal exposure | 1 | 310 | 430 | 440 | +2.3 | 430 | 0 | 430 | 440 | +2.3 | 440 | +2.3 | 425 | 330 | −22.3 | 310 | −27.1 |
| | 2 | 280 | 420 | 425 | +1.2 | 430 | +2.4 | 410 | 380 | −7.3 | 400 | −2.5 | 425 | 400 | −5.9 | 420 | −1.2 |
| | 3 | 245 | 400 | 390 | −2.5 | 390 | −2.5 | 430 | 410 | −4.6 | 420 | −2.3 | 380 | 320 | −15.8 | 310 | −18.4 |
| | 4 | 250 | 420 | 410 | −2.4 | 400 | −4.8 | 390 | 380 | −2.6 | 380 | −2.6 | 430 | 420 | −2.3 | 410 | −4.6 |
| | 5 | 265 | 395 | 400 | +1.3 | 400 | +1.3 | 405 | 400 | −1.2 | 400 | −1.2 | 410 | 380 | −7.3 | 380 | −7.3 |
| | Mean | 270.0 | 413 | 413 | −0.02 | 410 | −0.7 | 413 | 402 | −2.7 | 408 | −1.3 | 414 | 370 | −10.7 | 366 | −11.7 |
| | S.E. | 11.7 | 6.6 | 8.9 | 1.01 | 8.4 | 1.31 | 7.7 | 11.1 | 1.61 | 10.2 | 0.92 | 9.1 | 19.5 | 3.6 | 23.8 | 4.8 |
| | P | | | | NS | | NS | | | NS | | NS | | | 0.05 | | 0.05 |
| Tracheal exposure after atropine (1 mg/kg, i.v.) | 1 | 280 | 490 | 480 | −2.0 | 480 | −2.0 | 440 | 430 | −2.3 | 430 | −2.3 | 430 | 460 | +6.9 | 440 | +2.3 |
| | 2 | 270 | 460 | 480 | +4.3 | 470 | +2.2 | 410 | 410 | 0 | 410 | 0 | 420 | 410 | −2.4 | 395 | −5.9 |
| | 3 | 260 | 400 | 400 | 0 | 400 | 0 | 390 | 400 | +2.6 | 395 | +1.3 | 390 | 390 | 0 | 370 | −5.1 |
| | 4 | 265 | 460 | 450 | +2.2 | 460 | 0 | 430 | 430 | 0 | 440 | +2.3 | 460 | 460 | 0 | 450 | −2.2 |
| | 5 | 245 | 370 | 375 | +1.3 | 380 | +2.6 | 460 | 450 | −2.2 | 460 | 0 | 380 | 360 | −5.3 | 360 | −5.3 |
| | Mean | 264.0 | 436 | 437 | +1.2 | 438 | +0.6 | 426 | 424 | −0.4 | 427 | +0.3 | 416 | 416 | −0.2 | 403 | −3.2 |
| | S.E. | 5.8 | 22.0 | 21.3 | 1.1 | 20.1 | 0.8 | 12.0 | 8.7 | 0.9 | 11.3 | 0.8 | 14.3 | 19.6 | 2.0 | 18.1 | 1.5 |
| | P | | | | NS | | NS | | | NS | | NS | | | NS | | NS |

TABLE 3.9

**Electrocardiographic Effects of Isopropanol in Rats: P-R Interval (milliseconds)**

| | | | 3.9 mg/l | | | | | 19.5 mg/l | | | | | 97.5 mg/l | | | | |
|---|---|---|---|---|---|---|---|---|---|---|---|---|---|---|---|---|---|
| Procedure | Rat no. | Body wt. (g) | Initial | 3 min | %Δ | 5 min | %Δ | Initial | 3 min | %Δ | 5 min | %Δ | Initial | 3 min | %Δ | 5 min | %Δ |
| Nasal exposure | 1 | 310 | 42 | 42 | 0 | 40 | -4.8 | 40 | 41 | +2.5 | 42 | +5.0 | 40 | 40 | 0 | 41 | +2.5 |
| | 2 | 275 | 46 | 43 | -6.5 | 41 | -10.9 | 42 | 41 | -2.4 | 40 | -4.8 | 41 | 41 | 0 | 39 | -4.9 |
| | 3 | 290 | 47 | 47 | 0 | 48 | +2.1 | 48 | 48 | 0 | 50 | +4.2 | 48 | 48 | 0 | 49 | +2.1 |
| | 4 | 250 | 49 | 48 | -2.0 | 49 | 0 | 48 | 46 | -4.2 | 47 | -2.1 | 48 | 47 | -2.1 | 47 | -2.1 |
| | 5 | 265 | 43 | 43 | 0 | 43 | 0 | 44 | 44 | 0 | 43 | -2.3 | 44 | 44 | 0 | 44 | 0 |
| | Mean | 278 | 45.4 | 44.6 | -1.7 | 44.2 | -2.7 | 44.4 | 44.0 | -0.8 | 44.4 | 0 | 44.2 | 44.0 | -0.4 | 44.0 | -0.5 |
| | S.E. | 10.3 | 1.3 | 1.2 | 1.3 | 1.8 | 2.3 | 1.6 | 1.4 | 1.1 | 1.8 | 1.9 | 1.7 | 1.6 | 0.4 | 1.8 | 1.4 |
| | P | | | | NS | | NS | | | NS | | NS | | | NS | | NS |
| Nasal exposure after atropine (1 mg/kg, i.v.) | 1 | 270 | 48 | 46 | -4.2 | 47 | -2.1 | 49 | 50 | +2.0 | 51 | +3.9 | 51 | 51 | 0 | 52 | +2.0 |
| | 2 | 250 | 46 | 48 | +4.3 | 47 | +2.2 | 45 | 46 | +2.2 | 46 | +2.2 | 47 | 46 | -2.1 | 46 | -2.1 |
| | 3 | 255 | 52 | 50 | -3.8 | 50 | -3.8 | 50 | 51 | +2.0 | 50 | 0 | 52 | 51 | -1.9 | 49 | -5.8 |
| | 4 | 285 | 45 | 45 | 0 | 43 | -4.4 | 43 | 42 | -2.3 | 40 | -7.0 | 41 | 39 | -4.9 | 40 | -2.4 |
| | 5 | 300 | 48 | 49 | +2.1 | 49 | +2.1 | 44 | 45 | +2.2 | 44 | 0 | 46 | 45 | -2.2 | 44 | -4.3 |
| | Mean | 272.0 | 47.8 | 47.6 | -0.3 | 47.2 | -1.2 | 46.2 | 46.8 | +1.2 | 46.2 | -0.2 | 47.4 | 46.4 | -2.2 | 46.2 | -2.5 |
| | S.E. | 9.3 | 1.2 | 0.9 | 1.6 | 1.2 | 1.4 | 1.4 | 1.6 | 0.9 | 2.0 | 1.8 | 2.0 | 2.2 | 0.8 | 2.0 | 1.3 |
| | P | | | | NS | | NS | | | NS | | NS | | | 0.02 | | NS |
| Tracheal exposure | 1 | 310 | 48 | 46 | -4.2 | 46 | -4.2 | 47 | 46 | -2.1 | 45 | -4.2 | 46 | 47 | +2.1 | 47 | +2.1 |
| | 2 | 280 | 47 | 47 | 0 | 46 | -2.1 | 46 | 45 | -2.2 | 46 | 0 | 46 | 47 | +2.1 | 46 | 0 |
| | 3 | 245 | 46 | 45 | -2.2 | 43 | -6.5 | 43 | 43 | 0 | 45 | +4.6 | 43 | 40 | -7.0 | 40 | -7.0 |
| | 4 | 250 | 50 | 51 | +2.0 | 50 | 0 | 50 | 50 | 0 | 50 | 0 | 51 | 50 | -2.0 | 51 | 0 |
| | 5 | 265 | 49 | 48 | -2.0 | 45 | -8.2 | 45 | 44 | -2.2 | 44 | -2.2 | 45 | 46 | +2.2 | 46 | +2.2 |
| | Mean | 270.0 | 48.0 | 47.4 | -1.3 | 46.0 | -4.2 | 46.2 | 45.6 | -1.3 | 46.5 | -0.4 | 46.2 | 46.0 | -0.5 | 46.0 | -0.5 |
| | S.E. | 11.7 | 0.7 | 1.0 | 1.0 | 1.1 | 1.5 | 1.1 | 1.2 | 0.5 | 1.2 | 1.5 | 1.3 | 1.6 | 1.8 | 1.8 | 1.7 |
| | P | | | | NS | | 0.05 | | | NS | | NS | | | NS | | NS |
| Tracheal exposure after atropine (1 mg/kg, i.v.) | 1 | 280 | 42 | 43 | +2.3 | 42 | 0 | 42 | 41 | -2.4 | 40 | -4.8 | 40 | 41 | +2.5 | 40 | 0 |
| | 2 | 270 | 48 | 48 | 0 | 48 | 0 | 48 | 47 | -2.1 | 48 | 0 | 48 | 49 | +2.1 | 47 | -2.1 |
| | 3 | 260 | 50 | 50 | 0 | 51 | +2.0 | 50 | 49 | -2.0 | 50 | 0 | 51 | 50 | -2.0 | 50 | -2.0 |
| | 4 | 265 | 49 | 48 | -2.0 | 49 | 0 | 49 | 49 | 0 | 49 | 0 | 49 | 48 | -2.0 | 48 | -2.0 |
| | 5 | 245 | 42 | 41 | -2.4 | 41 | -2.4 | 40 | 40 | 0 | 41 | +2.5 | 41 | 41 | 0 | 41 | 0 |
| | Mean | 264 | 46.2 | 46.0 | -0.4 | 46.2 | -0.1 | 45.8 | 45.2 | -1.3 | 45.6 | -0.5 | 45.8 | 45.8 | +0.12 | 45.2 | -1.2 |
| | S.E. | 5.8 | 1.7 | 1.7 | 0.8 | 2.0 | 0.7 | 2.0 | 1.9 | 0.5 | 2.1 | 1.2 | 2.2 | 2.0 | 0.9 | 2.0 | 0.5 |
| | P | | | | NS | | NS | | | NS | | NS | | | NS | | NS |

response to the same concentrations of isopropanol were -1.5% and -10.0% in unexposed rats and -4.2% and -22.9% in exposed rats. Here the loss of compliance induced by the alcohol was more pronounced in the animals previously exposed to it. Respiratory rate changes did not differ significantly between the two groups. The tidal volume reductions produced by the most concentrated of the gas mixtures were comparable between the two groups, while the intermediate concentration (19.5 mg/l) evoked a stronger response in the exposed animals, compared to the previously untreated control animals (Table 3.10).

A second stimulus was tested on another group of five rats exposed daily for 6 days to isopropanol (400 ppm) as described previously. These and five control animals were exposed for 1 min to a gas mixture of 8.4% oxygen and 91.6% nitrogen. Pulmonary resistance, pulmonary compliance, respiratory rate, and tidal volume were computed by the pulmonary mechanics computer. In each parameter, the maximum response to the hypoxia was recorded. It is clear from Table 3.11 that the 6-day subacute exposures to isopropanol vapor received by the second group of rats impeded the pulmonary compensatory response to hypoxia, compared to the response of the control animals with no previous exposure to the solvent.

*5. Effects of Subacute Inhalation on Liver*

In order to assess the hepatotoxic effects of acute and subacute exposures to isopropanol vapor, serum glutamate-pyruvate transaminase (SGPT) activity was determined in three groups of differently treated rats. In all trials, blood was withdrawn by cardiac puncture from animals under light anaesthesia. SGPT activity was determined by a standard spectrophotometric assay. The groups of five animals were pretreated in the following manner: (1) control rats – no prior treatment; (2) subacute exposure to 400 ppm (0.098 mg/l) isopropanol on 6 consecutive days, blood withdrawn on seventh day; (3) subacute exposure to isopropanol for 6 days followed on the seventh day with three consecutive 3-min exposures to isopropanol gas mixtures of 3.9, 19.5, and 97.5 mg/l, blood withdrawn immediately after exposures. Determination of SGPT activity in the three groups of animals produced the following results:

| Treatment | SGPT Activity |
|---|---|
| Controls | 13.4 ± 1.2 I.U. |
| Subacute exposure | 15.3 ± 0.8 I.U. |
| Subacute and acute exposure | 14.3 ± 0.4 I.U. |

No significant differences between the three values are evident and all fall within the normal range of enzyme activity in rat serum (13.0 to 17.0 I.U.).

The control and exposed rats described above were finally exsanguinated by cutting the abdominal aorta. Segments of livers and testes were placed in buffered neutral formalin within 5 min after death. Trachea, bronchi, and lungs were removed intact, the lungs allowed to collapse by gravity, then expanded by injecting fixative via the trachea and dropped into the fixative, also within 5 min after death. Tissues for comparisons were taken from five negative control male rats of approximately the same age. Tissues were embedded in paraffin, cut at 5 $\mu$m and stained by hematoxylin and eosin. The only difference detected by the post-mortem examination between rats that had inhaled isopropanol vapor and the controls was in the numbers of mitotic figures found in the hepatocytes. Table 3.12 illustrates this difference through counts of mitotic figures per 1000 hepatocytes in ten random microscopic fields as seen in a grid at 100 magnification. Mitotic figures were approximately four times more numerous among hepatocytes of the controls.

*6. Effects of Intravenous Injection on the Heart*

Three rats anesthetized with 65 mg/kg sodium pentobarbital injected intraperitoneally were used in this study.

The trachea was cannulated and the rat was artificially ventilated with a Starling pump, which delivered a volume of 5 ml of air at a rate of 60/min. The left common carotid was cannulated with polyethylene tubing, which was attached to a Statham transducer in order to measure blood pressure. A steel needle was inserted into the right femoral vein for the injection of isopropanol.

The chest was opened at the tip of the xyphoid process. A chest wall flap was created by cutting across the anterior border of the diaphragm and superiorly to the third or fourth intercostal space from each end of the diaphragmatic incision. A

TABLE 3.10

**Respiratory Effects of Isopropanol: Subacute Inhalation**

| Procedure | Rat no. | Body wt. (g) | Pulmonary resistance ($cm\ H_2O/ml/sec$) | | | Pulmonary compliance ($ml/cm\ H_2O$) | | | Respiratory rate (/min) | | | Tidal volume (ml) | | |
|---|---|---|---|---|---|---|---|---|---|---|---|---|---|---|
| | | | Before | 3 min | %Δ | Before | 3 min | %Δ | Before | 3 min | %Δ | Before | 3 min | %Δ |
| Isopropanol | 1 | 210 | 0.193 | 0.186 | −3.4 | 0.788 | 0.800 | +1.6 | 87.5 | 87.5 | 0.0 | 1.450 | 1.420 | −2.2 |
| 3 min | 2 | 225 | 0.140 | 0.140 | 0 | 0.450 | 0.463 | +2.8 | 97.5 | 102.5 | +5.1 | 1.360 | 1.280 | −5.9 |
| 3.9 mg/l | 3 | 230 | 0.160 | 0.166 | +4.2 | 0.350 | 0.388 | +10.7 | 112.5 | 107.5 | −4.4 | 1.440 | 1.438 | +3.3 |
| | 4 | 210 | 0.166 | 0.166 | 0 | 0.288 | 0.294 | +2.2 | 85.0 | 83.5 | −1.8 | 1.200 | 1.200 | 0 |
| | 5 | 222 | 0.180 | 0.173 | −3.7 | 0.300 | 0.313 | +4.2 | 90.0 | 88.8 | −1.4 | 1.392 | 1.360 | −2.3 |
| | Mean | 219.4 | 0.168 | 0.166 | −0.6 | 0.435 | 0.452 | +4.3 | 94.5 | 94.0 | −0.5 | 1.368 | 1.339 | −1.42 |
| | S.E. | 4.0 | 0.009 | 0.007 | 1.4 | 0.093 | 0.092 | 1.7 | 5.0 | 1.7 | 1.6 | 0.045 | 0.044 | 1.5 |
| | P | | | | NS | | | NS | | | NS | | | NS |
| Isopropanol | 1 | 210 | 0.186 | 0.203 | +9.3 | 0.810 | 0.750 | −7.4 | 90.0 | 90.0 | 0 | 1.392 | 1.300 | −6.6 |
| 3 min | 2 | 225 | 0.140 | 0.140 | 0 | 0.450 | 0.469 | −4.2 | 105.0 | 114.0 | +8.6 | 1.240 | 1.136 | −8.4 |
| 19.5 mg/l | 3 | 230 | 0.166 | 0.190 | +14.0 | 0.375 | 0.363 | −3.3 | 107.5 | 118.5 | +10.2 | 1.448 | 1.200 | −17.1 |
| | 4 | 210 | 0.160 | 0.170 | +6.3 | 0.288 | 0.281 | −2.2 | 80.0 | 77.5 | −3.1 | 1.180 | 1.120 | −5.1 |
| | 5 | 222 | 0.173 | 0.186 | +7.5 | 0.338 | 0.325 | −3.7 | 91.3 | 87.5 | −4.2 | 1.320 | 1.200 | −9.1 |
| | Mean | 219.4 | 0.165 | 0.178 | +7.4 | 0.452 | 0.438 | −4.2 | 94.8 | 97.5 | +4.0 | 1.316 | 1.191 | −9.3 |
| | S.E. | 4.0 | 0.008 | 0.011 | 2.3 | 0.093 | 0.084 | 0.9 | 5.1 | 8.0 | 2.5 | 0.049 | 0.032 | 2.0 |
| | P | | | | 0.05 | | | NS | | | NS | | | 0.02 |
| Isopropanol | 1 | 210 | 0.186 | 0.306 | +64.3 | 0.763 | 0.588 | −22.9 | 92.5 | 80.0 | −13.5 | 1.355 | 1.300 | −4.1 |
| 3 min | 2 | 225 | 0.133 | 0.173 | +30.0 | 0.438 | 0.350 | −20.0 | 117.5 | 97.5 | −17.0 | 1.256 | 1.120 | −10.8 |
| 97.5 mg/l | 3 | 230 | 0.180 | 0.207 | +15.6 | 0.350 | 0.258 | −26.4 | 117.5 | 105.0 | −10.6 | 1.380 | 1.240 | −10.1 |
| | 4 | 210 | 0.173 | 0.203 | +17.3 | 0.294 | 0.213 | −27.6 | 78.8 | 73.8 | −6.3 | 1.240 | 1.120 | −9.7 |
| | 5 | 222 | 0.180 | 0.236 | +31.4 | 0.313 | 0.258 | −17.6 | 90.0 | 81.3 | −9.7 | 1.420 | 1.300 | −8.5 |
| | Mean | 219.4 | 0.170 | 0.225 | +31.7 | 0.432 | 0.333 | −22.9 | 99.3 | 87.5 | −11.4 | 1.330 | 1.216 | −8.6 |
| | S.E. | 4.0 | 0.009 | 0.023 | 8.7 | 0.087 | 0.067 | 1.9 | 7.8 | 5.9 | 1.8 | 0.035 | 0.040 | 1.2 |
| | P | | | | 0.05 | | | 0.02 | | | 0.01 | | | NS |

TABLE 3.11

**Respiratory Effects of Hypoxia After Subacute Inhalation of Isopropanol**

| Procedure | Rat no. | Body wt. (g) | Pulmonary resistance ($cm H_2O/ml/sec$) | | | Pulmonary compliance ($ml/cm H_2O$) | | | Respiratory rate (/min) | | | Tidal volume (ml) | | |
|---|---|---|---|---|---|---|---|---|---|---|---|---|---|---|
| | | | Before | 3 min | %Δ | Before | 3 min | %Δ | Before | 3 min | %Δ | Before | 3 min | %Δ |
| 8.4% $O_2$ 1 min | 1 | 240 | 0.126 | 0.113 | −10.5 | 0.333 | 0.450 | +35.3 | 85.0 | 115.0 | +35.3 | 1.760 | 2.424 | +37.7 |
| | 2 | 250 | 0.156 | 0.146 | −6.4 | 0.308 | 0.438 | +42.3 | 115.0 | 165.0 | +43.5 | 1.384 | 2.140 | +54.6 |
| | 3 | 235 | 0.162 | 0.145 | −10.7 | 0.375 | 0.520 | +38.7 | 92.5 | 131.0 | +41.6 | 1.520 | 2.600 | +84.2 |
| | 4 | 220 | 0.372 | 0.319 | −14.3 | 0.425 | 0.593 | +39.4 | 132.5 | 162.5 | +22.6 | 1.144 | 1.620 | +41.6 |
| | 5 | 242 | 0.146 | 0.126 | −15.2 | 0.533 | 0.788 | +47.9 | 97.5 | 132.5 | +35.9 | 1.520 | 3.360 | +121.1 |
| | Mean | 237.4 | 0.192 | 0.170 | −11.4 | 0.395 | 0.558 | +40.7 | 104.5 | 141.2 | +35.8 | 1.466 | 2.429 | +67.8 |
| | S.E. | 5.0 | 0.045 | 0.038 | 1.6 | 0.040 | 0.064 | 2.1 | 8.6 | 9.7 | 3.6 | 0.100 | 0.286 | 15.6 |
| | P | | | | NS | | | 0.01 | | | 0.001 | | | 0.02 |
| 8.4% $O_2$ 1 min, after 6 days × 1 hr exposure to 400 ppm | 1 | 239 | 0.163 | 0.173 | +6.1 | 0.400 | 0.533 | +33.1 | 130.0 | 162.5 | +25.0 | 1.480 | 2.480 | +67.6 |
| | 2 | 255 | 0.213 | 0.193 | −9.4 | 0.258 | 0.300 | +16.5 | 120.0 | 145.0 | +20.8 | 1.120 | 1.440 | +28.6 |
| | 3 | 230 | 0.295 | 0.246 | −16.7 | 0.270 | 0.318 | +17.6 | 122.5 | 153.8 | +25.5 | 1.680 | 2.192 | +9.6 |
| | 4 | 246 | 0.160 | 0.154 | −3.3 | 0.313 | 0.525 | +68.0 | 130.0 | 180.0 | +38.5 | 1.320 | 2.104 | +59.4 |
| | 5 | 225 | 0.210 | 0.186 | −11.4 | 0.350 | 0.413 | +17.8 | 127.0 | 147.0 | +15.7 | 1.400 | 1.560 | +11.4 |
| | Mean | 239.0 | 0.208 | 0.190 | −6.9 | 0.318 | 0.418 | +30.6 | 125.9 | 157.7 | +25.1 | 1.400 | 1.955 | +35.3 |
| | S.E. | 5.4 | 0.024 | 0.015 | 3.9 | 0.026 | 0.049 | 9.8 | 2.0 | 6.4 | 3.8 | 0.092 | 0.197 | 12.0 |
| | P | | | | NS | | | 0.05 | | | 0.01 | | | 0.05 |

TABLE 3.12

**Counts Within a 10 × 10 Grid (× 100) of Mitoses in Hepatocytes of Rats After Isopropanol Vapor Inhalation Compared to Controls**

| | Mitoses/10,000 cells | | | | | | | | | |
|---|---|---|---|---|---|---|---|---|---|---|
| | Isopropanol exposed | | | | | Control rats | | | | |
| Microscopic field | Rat 1 | Rat 2 | Rat 3 | Rat 4 | Rat 5 | Rat 1 | Rat 2 | Rat 3 | Rat 4 | Rat 5 |
| 1 | 2 | 0 | 0 | 1 | 0 | 2 | 4 | 2 | 4 | 3 |
| 2 | 2 | 0 | 2 | 0 | 0 | 1 | 1 | 3 | 4 | 2 |
| 3 | 3 | 1 | 2 | 2 | 0 | 2 | 4 | 3 | 0 | 4 |
| 4 | 1 | 1 | 1 | 0 | 0 | 4 | 5 | 0 | 3 | 1 |
| 5 | 2 | 0 | 0 | 0 | 0 | 2 | 5 | 2 | 1 | 4 |
| 6 | 3 | 1 | 0 | 0 | 0 | 3 | 3 | 2 | 5 | 2 |
| 7 | 0 | 0 | 0 | 0 | 0 | 0 | 1 | 6 | 2 | 1 |
| 8 | 1 | 1 | 0 | 0 | 0 | 4 | 0 | 3 | 0 | 1 |
| 9 | 2 | 0 | 0 | 0 | 0 | 0 | 2 | 0 | 2 | 4 |
| 10 | 2 | 1 | 0 | 0 | 0 | 2 | 2 | 3 | 4 | 2 |
| Total | 18 | 5 | 5 | 3 | 0 | 20 | 27 | 24 | 25 | 24 |
| Mean | | | 0.62 | | | | | 2.4 | | |

120-ohm rat-size strain gauge arch was sutured to the left ventricle and connected to a force-displacement transducer. The magnitude and number of contractions per unit time were recorded on a precalibrated Sanborn 150 oscillograph.

An intravenously injected dose of 1.0 ml/kg isopropanol produced a mean reduction of 45.0% in myocardial contractile force and a concomitant mean reduction of 38.9% in arterial blood pressure (Table 3.13). The latter change was a more transient one, as autonomic reflexes act to maintain blood pressure as cardiac output falls. The mechanism by which the solvent depresses myocardial contractility is not deducible from this study.

## C. Discussion

The observations relating to respiration are new and represent the first reported investigation of functional changes. It is apparent that the current TLV of 400 ppm (about 0.1 mg/l) has no acute and immediate effects on the respiratory system of the rat. Even levels about 40 times the TLV (3.9 mg/l) do not influence lung mechanics. Supersaturated isopropanol-air mixtures (97.5 mg/l) were necessary to produce bronchoconstriction which was blocked by atropine, and to cause a potentiation of increase in tidal volume and respiratory rate by isoproterenol. The interaction between atropine and isopropanol can be interpreted to mean that the bronchoconstriction produced by isopropanol is mediated by cholinergic receptors. However, the potentiation of isoproterenol effects cannot be readily explained since isopropanol is not known to interfere with the metabolism of sympathomimetic amines.

The threshold concentration for depression of respiratory minute volume is 19.5 mg/l. The depressant effect of nasal inhalation is more intense than tracheal administration. The difference indicates that passage of isopropanol through the nasal cavity and nasopharynx cause respiratory depression representing a response to irritation of the mucosal receptors. Pretreatment with atropine eliminates the respiratory response to isopropanol. This interaction cannot be readily explained because the mucosal receptors as well as the innervation of the respiratory muscles are not known to be susceptible to blockade by atropine. It is possible that isopropanol is acting on cholinergic receptors in the respiratory reflex arc which have not been previously detected.

The effects of isopropanol on the cardiovascular system consist of a fall in systemic arterial blood pressure and slowing of the heart rate seen only when high concentration of isopropanol (97.5 mg/l) is inhaled. It was not technically possible to measure myocardial contractility during inhalation of the solvent. The intravenous injection caused depression of contractile force.

The few rats that were subacutely exposed to the Threshold Limit Value of 400 ppm (0.098

mg/l) for 1 hr a day for 6 days showed a change in the respiratory system. There was a hyperresponsiveness to the acute effects of isopropanol and hyporesponsiveness to hypoxia. These results are new and need to be investigated further. The same animals do not show disturbance in liver function although there was a decrease in mitotic figures in the hepatic cells.

### D. Summary

Isopropanol-air mixtures in supersaturated concentration (97.5 mg/l) caused bronchoconstriction, respiratory minute volume depression, hypotension, and bradycardia in rats. Daily 1-hr exposure to Threshold Limit levels of isopropanol (0.098 mg/l) caused a change in responsiveness of the airways.

TABLE 3.13

**Cardiovascular Effect of Intravenous Injection of Isopropanol in Rats**

| Rat no. | Body wt. (g) | Myocardial force (g) Before | Myocardial force (g) After | Myocardial force (g) %Δ | Arterial pressure (mm Hg) Before | Arterial pressure (mm Hg) After | Arterial pressure (mm Hg) %Δ |
|---|---|---|---|---|---|---|---|
| 1 | 220 | 2.18 | 1.25 | −42.7 | 113 | 73 | −35.4 |
| 2 | 245 | 3.13 | 1.87 | −39.9 | 125 | 72 | −42.4 |
| 3 | 228 | 3.47 | 1.65 | −52.4 | 90 | 55 | −38.9 |
| Mean | 231 | 2.93 | 1.59 | −45.0 | 109.3 | 66.7 | −38.9 |
| SE | 7.4 | 0.39 | 0.18 | 3.8 | 10.3 | 6.0 | 2.0 |

## Chapter 4
## HEMODYNAMIC EFFECTS OF ISOPROPANOL IN THE ANESTHETIZED DOG

Although the rat preparation could effectively show the effects of isopropanol on the respiratory system, it was not technically feasible to examine the corresponding responses of the circulatory system (Chapter 3). The dog preparation used for fluorinated propellants[1] and chlorinated solvents[2] has been applied to the study of isopropanol. The results described below represent the first reported investigation of the hemodynamic effects of isopropanol. Two solvents were tested in the following concentrations: isopropanol 0.5% (1.22 mg/l), 1.0% (2.45 mg/l), 2.5% (6.12 mg/l), 5.0% (12.25 mg/l), and 7.5% (18.37 mg/l); and trichloroethylene 0.01% (0.535 mg/l).

### A. Methods

The experiments were performed on adult mongrel dogs of either sex, weighing 18 to 24 kg (average 20.8) which were anesthetized by intravenous administration of 30 to 35 mg/kg pentobarbital sodium. Tracheotomy was then carried out and artificial respiration was maintained by Starling Ideal respirator. Aortic blood pressure was measured by inserting a catheter attached to a Statham P23 AA pressure transducer into the right carotid artery. The chest was opened by an incision in the fourth intercostal space of the left side. The pericardium was opened and the pulmonary artery was dissected free from the aorta. An electromagnetic flow probe with an internal diameter of approximately 10% less than the pulmonary artery was fitted around the latter. The pulmonary arterial blood flow was thus measured using a Statham SP 2202 electromagnetic flowmeter. Left atrial pressure and pulmonary arterial pressure were also measured by introducing catheters connected to a P23 BB Statham pressure transducer into the left atrial appendage and into the pulmonary artery to the left lower lobe.

The chest was also entered in the right side at the sixth intercostal space and a catheter, attached to a Statham P23 AA pressure transducer, was introduced into the left ventricular lumen to measure the left ventricular pressure. The signal was fed to a differentiating circuit (8814A HP Derivative computer) and the left ventricular dp/dt was recorded. All parameters were recorded on a six-channel Sanborn 7700 recorder.

About 30 min elapsed before the administration of isopropanol vapors, during which the preparation was allowed to stabilize. Various concentrations of isopropanol in air were then administered in succession, via the inlet of the respirator, each for 5 min. Various concentrations of isopropanol were prepared by volatilizing the appropriate amount of the solvent, calculated to give the appropriate volume of vapors at standard pressure and 25°C, as previously described.[1]

The minimal effective concentration of isopropanol was determined and a subthreshold concentration (usually 50% of the threshold concentration) was tested in the same preparation. Trichloroethylene (0.01%, a subthreshold concentration that was previously determined in a similar preparation)[2] alone, or together with the subthreshold concentration of isopropanol, was administered for 5 min to find out any possible interaction between isopropanol and trichloroethylene insofar as the hemodynamic and myocardial effects are concerned. The sequence of administration of isopropanol, trichloroethylene, or the mixture of them, is altered from one experiment to the other.

All results are represented as mean ±SE. Student's t-test modified for paired replicates was used to compare data obtained from the same animal. To test for the significance of the difference between the effects of isopropanol, trichloroethylene, and/or the mixture of both the Student's t-test was used. In all cases, P values of $<0.05$ were considered significant.

### B. Calculations and Abbreviations of Hemodynamic Measurements

MPAP: Mean pulmonary arterial pressure in cm $H_2O$, measured from the pulmonary artery to the left lower lobe.

MLAP: Mean left atrial pressure in cm $H_2O$, measured from the left atrium.

EMPAP: Effective mean pulmonary arterial pressure in cm $H_2O$, mean pulmonary arterial pressure minus mean left atrial pressure.

LVP: Left ventricular pressure in mm Hg, measured from a catheter in the left ventricular cavity inserted through the cardiac apex.

LVEDP: Left ventricular end-diastolic pressure in mm Hg, measured from the left ventricular pressure.

dp/dt: Maximal rate of rise of left ventricular pressure in mm Hg/sec derived from the left ventricular pressure with a derivative computer, type HP 8814A.

MAP: Mean aortic pressure in mm Hg, measured from the right carotid artery.

MPAF: Mean pulmonary arterial flow in ml/min, measured with a Statham electromagnetic flow probe around the main pulmonary artery, using a Statham electromagnetic flowmeter.

HR: Heart rate, in beats/min, computed from aortic pressure waves.

Systemic vascular resistance: In dynes·sec/$cm^5$; the quotient of mean aortic pressure minus left ventricular end-diastolic pressure in dynes/$cm^2$ and of mean pulmonary arterial flow in ml/sec.

Pulmonary vascular resistance: In dynes·sec/$cm^5$; the quotient of mean pulmonary arterial pressure minus mean left atrial pressure in dynes/$cm^5$ and of mean pulmonary arterial flow in ml/sec.

SV: Stroke volume: in ml, computed from cardiac output and heart rate.

SW: Stroke work: in gram·meter (g·m), computed from stroke volume and aortic pressure.

## C. Results

The results relating to isopropanol are described first, followed by those of the interaction between isopropanol and trichloroethylene.

### *1. Isopropanol*

Three stages appeared during the inhalation of progressively increasing concentrations of isopropanol (Table 4.1). The order of the three stages is as follows:

**Myocardial depression** – Myocardial depression (5.7%) occurs first simultaneously with acceleration of heart rate (1.7%) and increases in pulmonary vascular resistance. The minimal threshold concentration is 1.0% isopropanol. The myocardial and hemodynamic effects of various concentrations of isopropanol, from a typical experiment, are shown in Figure 4.1. The effect of isopropanol on myocardial contractility as well as on other parameters is depicted in Figure 4.2.

**Reduced mean pulmonary arterial flow** – Reduced mean pulmonary arterial flow (5.5%) or reduced output with reduction in stroke volume (8.5%) and stroke work (10.3%) is seen during inhalation of 2.5% isopropanol.

**Systemic hypotension** – Systemic hypotension (4.4%) without a change in systemic and pulmonary vascular resistance indicates that this last stage seen with 7.5% is a further manifestation of the earlier stages of myocardial depression and reduced cardiac output.

### *2. Interaction between Isopropanol and Trichloroethylene*

The five dogs comprising the group that were subjected to isopropanol inhalation had varying threshold concentrations for appearance of signs of myocardial depression. They ranged from 0.5 to 2.5%, with a mean concentration of 1.4%. The mean level for the concentration that did not elicit an effect, i.e., subthreshold concentration, was 0.63 ± 0.13%. Each dog also was tested for subthreshold concentrations of trichloroethylene (0.01%) and then a mixture of both isopropanol and trichloroethylene was administered. The results summarized in Table 4.2 show a lack of potentiation of the individual hemodynamic effect of isopropanol and trichloroethylene. The reduction in stroke volume and of stroke work with the combination of the two solvents represents an additive effect of iospropanol and trichloroethylene separately (Figure 4.3).

## D. Discussion

Isopropanol inhaled in a concentration of 1.0% causes depression of myocardial contractility. This threshold dose for isopropanol is higher than 0.25% for methyl chloroform and 0.05% for trichloroethylene reported previously using a similar heart preparation.[2] The results collectively show that isopropanol has a lower level of cardiotoxicity than the chlorinated solvents. When isopropanol is used concurrently with trichloroethylene, there is no potentiation of cardiotoxicity although a summation of their individual effects may appear.

The experiments reported above are the first hemodynamic investigation of isopropanol. The essential features are as follows: (1) Isopropanol depresses myocardial contractility at a concentration of 1.0%. To further elucidate the nature of the myocardial effects of isopropanol, its effect on left ventricular dp/dt, stroke work and left ventricular end diastolic pressure are shown in Figure 4.4. The inhalation of isopropanol brought about a decrease of 2.8%, 5.7%, 7.4%, 11.7%, and 12.9% in left ventricular dp/dt with concentrations of 0.5, 1.0, 2.5, 5.0, and 7.5%, respectively. (2) The concentration of isopropanol that reduced cardiac output is 2.5%. It seems most probable that the decrease in cardiac output might be due mainly to depression in myocardial contractility and decrease in systolic ejection. (3) The concentration of isopropanol that reduced systemic arterial blood pressure is 7.5%, which is three times higher than the concentration that depressed cardiac function. (4) The inhalation of isopropanol as high as 10% did not influence pulmonary vascular resistance.

**E. Summary**

The inhalation of isopropanol elicits the following hemodynamic effects in the anesthetized dog: depression of myocardial contractility with a 1.0% concentration (2.45 mg/l); reduction in cardiac output with 2.5% (6.12 mg/l); and systemic hypotension with 7.5% (18.37 mg/l). There is no potentiation of the cardiac effects of subthreshold concentrations of isopropanol and trichloroethylene.

TABLE 4.1

**Hemodynamic Effects of Inhalation of Progressively Increasing Concentrations of Isopropanol in Anesthetized Dogs***

| | MPAP cm $H_2O$ | MLAP cm $H_2O$ | EMPAP cm $H_2O$ | LVP mmHg | LVEDP mmHg | dp/dt mmHg/sec | MAP mmHg | MPAF ml/min | HR beats/min | Vascular resistance dynes·sec/$cm^5$ | | Stroke vol ml | Stroke work g·m |
|---|---|---|---|---|---|---|---|---|---|---|---|---|---|
| | | | | | | | | | | Pulmonary | Systemic | | |
| Control | 33.0 ± 5.38 | 8.8 ± 0.77 | 24.9 ± 4.90 | 136.8 ± 10.76 | 1.3 ± 0.88 | 3792 ± 320.6 | 121.5 ± 10.91 | 1492 ± 52.3 | 175 ± 16.1 | 1014 ± 229.7 | 6488 ± 634.7 | 8.9 ± 0.85 | 15.3 ± 1.62 |
| 0.5% | 33.0 ± 5.32 | 8.7 ± 0.88 | 25.1 ± 4.86 | 136 ± 10.59 | 1.3 ± 0.88 | 3688 ± 333.5 | 121.5 ± 10.48 | 1467 ± 52.7 | 175 ± 16.1 | 1050 ± 235.6 | 6605 ± 658.1 | 8.7 ± 0.87 | 14.9 ± 1.60 |
| isopropanol | 0 ± 0.38 | −0.1 ± 0.18 | +0.2 ± 0.49 | −0.8 ± 1.81 | 0 ± 0 | −104 ± 81.76 | 0 ± 1.88 | −25 ± 21.4 | +0.5 ± 1.06 | +36 ± 17.7 | +117 ± 72.3 | −0.2 ± 0.12 | −0.4 ± 0.43 |
| | NS | NS | NS | NS | NS | NS | NS | NS | NS | NS | NS | NS | NS |
| 1.0% | 32.6 ± 5.39 | 8.6 ± 0.93 | 24.8 ± 4.91 | 134.8 ± 10.81 | 1.3 ± 0.88 | 3583 ± 331.8 | 120.2 ± 11.18 | 1442 ± 55.4 | 178 ± 15.9 | 1063 ± 243.3 | 6658 ± 728.4 | 8.5 ± 0.89 | 14.4 ± 1.78 |
| | −0.4 ± 0.46 | −0.2 ± 0.25 | −0.04 ± 0.51 | −2.0 ± 2.63 | 0 ± 0 | −208 ± 69.72 | −1.3 ± 2.32 | −50 ± 31.6 | +2.7 ± 0.67 | +49 ± 17.1 | +170 ± 124.7 | −0.4 ± 0.17 | −0.9 ± 0.51 |
| | NS | NS | NS | NS | NS | 0.05 | NS | NS | 0.01 | 0.05 | NS | NS | NS |
| 2.5% | 32.9 ± 5.47 | 9.1 ± 1.13 | 24.7 ± 4.89 | 134 ± 11.16 | 1.4 ± 1.80 | 3479 ± 343.7 | 119.3 ± 11.51 | 1408 ± 53.9 | 180 ± 16.4 | 1080 ± 249.9 | 6776 ± 784.4 | 8.2 ± 0.87 | 13.8 ± 1.64 |
| | −0.2 ± 0.50 | +0.3 ± 0.42 | −0.16 ± 0.53 | −2.8 ± 2.50 | +0.1 ± 0 | −313 ± 100.8 | −2.2 ± 1.82 | −84 ± 30.7 | +5.0 ± 0.85 | +66 ± 27.1 | +288 ± 214.3 | −0.7 ± 0.19 | −1.5 ± 0.35 |
| | NS | NS | NS | NS | NS | 0.02 | NS | 0.05 | 0.01 | NS | NS | 0.02 | 0.01 |
| 5.0% | 32.9 ± 5.69 | 9.3 ± 1.14 | 24.5 ± 5.00 | 132.8 ± 11.89 | 1.7 ± 1.05 | 3375 ± 380.5 | 118.8 ± 11.90 | 1392 ± 37.5 | 183 ± 16.3 | 1061 ± 239.5 | 6760 ± 731.4 | 8.0 ± 0.76 | 13.3 ± 1.54 |
| | −0.1 ± 0.43 | +0.5 ± 0.43 | −0.38 ± 0.51 | −4 ± 2.72 | +0.3 ± 0.3 | −417 ± 127.8 | −2.7 ± 2.19 | −100 ± 22.4 | +8.0 ± 1.79 | +47 ± 32.9 | +272 ± 215.9 | −1.0 ± 0.23 | −2.0 ± 0.31 |
| | NS | NS | NS | NS | NS | 0.02 | NS | 0.01 | 0.01 | NS | NS | 0.01 | 0.01 |
| 7.5% | 32.0 ± 5.09 | 9.5 ± 1.10 | 23.2 ± 4.35 | 129.8 ± 11.64 | 2.5 ± 1.12 | 3333 ± 389.5 | 116.3 ± 11.05 | 1342 ± 55.4 | 176 ± 15.9 | 1030 ± 227.8 | 6875 ± 799.8 | 7.9 ± 0.46 | 12.2 ± 1.21 |
| | −1.0 ± 0.70 | +0.7 ± 0.45 | −1.64 ± 1.17 | −7.0 ± 2.43 | +1.2 ± 0.83 | −458 ± 139.4 | −5.2 ± 1.42 | −150 ± 40.8 | 10.6 ± 3.35 | +16 ± 64.1 | +387 ± 285.9 | −1.5 ± 0.57 | −3.2 ± 1.08 |
| | NS | NS | NS | 0.05 | NS | 0.02 | 0.02 | 0.02 | 0.02 | NS | NS | 0.05 | 0.05 |

*Each group of numbers includes mean ± standard error for five dogs, mean difference ± standard error, and the significance level.

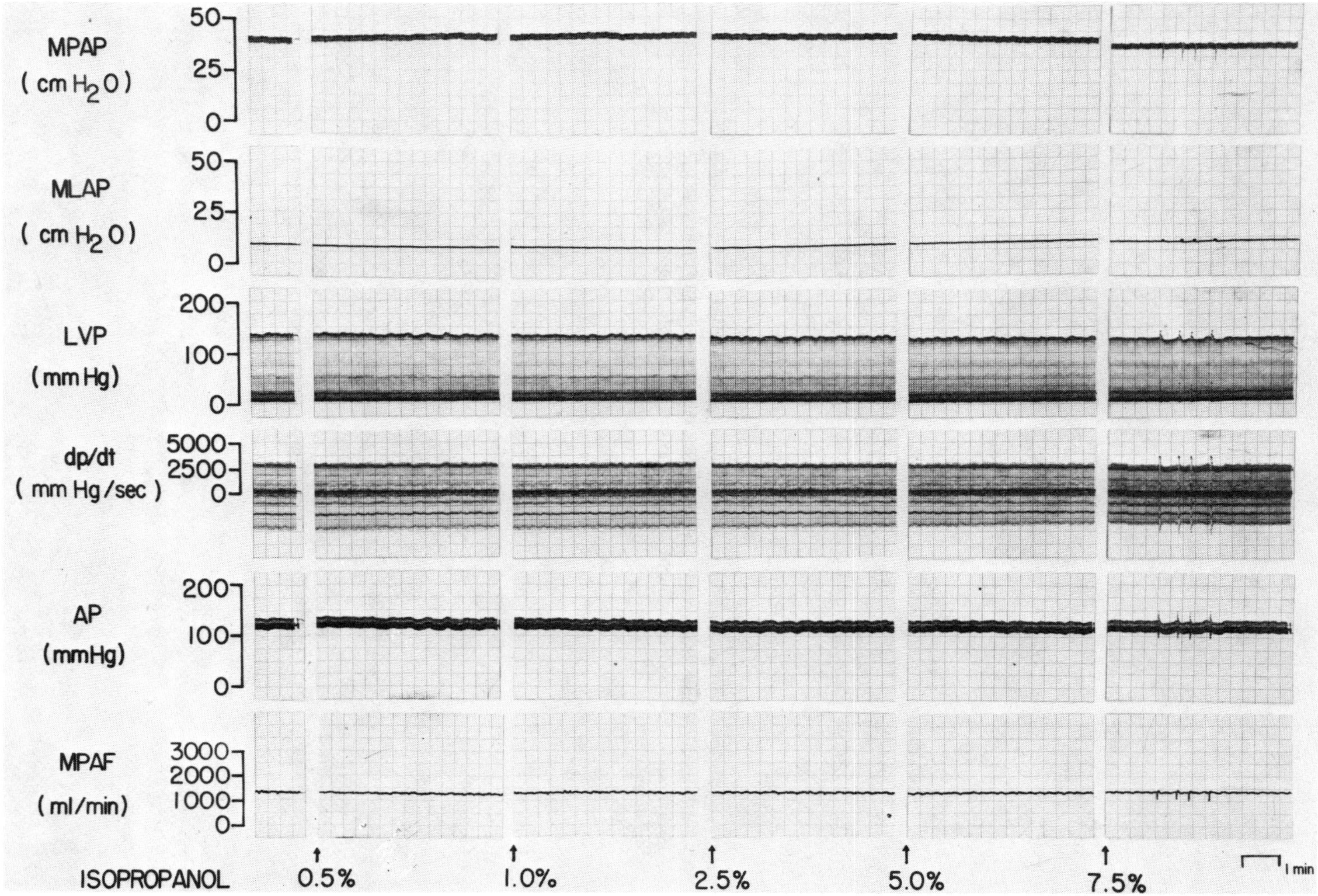


FIGURE 4.1. The effect of various, progressively increasing concentrations of isopropanol on the myocardium and hemodynamics of the open-chest dog preparation. Abbreviations are explained in the text.

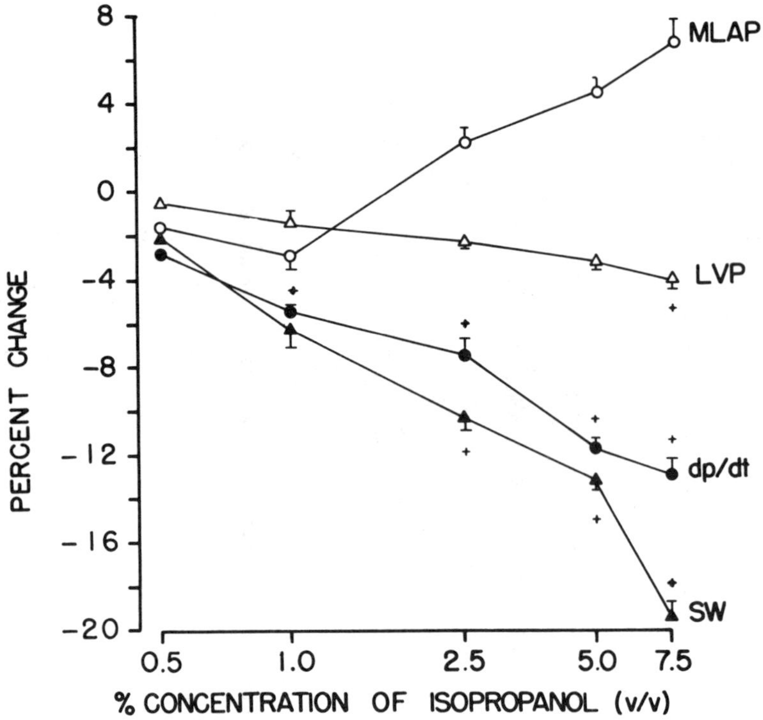


FIGURE 4.2. Concentration-response curve of isopropanol showing the changes in mean left atrial pressure (MLAP), left ventricular pressure (LVP), left ventricular dp/dt and stroke work (SW). Note that the first significant change was in myocardial contractility brought about by as low a concentration as 1%. + = significant change.

TABLE 4.2

**Interaction of Subthreshold Concentrations of Isopropanol and Trichloroethylene***

| | Isopropanol subthreshold (0.63 ± 0.13%) | | Trichloroethylene subthreshold (0.01%) | | Isopropanol and trichloroethylene | |
|---|---|---|---|---|---|---|
| | C | E | C | E | C | E |
| MPAP | 33.3 | 33.0 | 33.3 | 33.5 | 33.3 | 32.3 |
| (cm $H_2O$) | −0.2 | ±0.24 | +0.2 | ±0.59 | −1.0 | ±0.47 |
| P | NS | | NS | | NS | |
| MLAP | 8.6 | 8.7 | 8.3 | 8.4 | 8.2 | 7.8 |
| (cm $H_2O$) | +0.1 | ±0.10 | +0.1 | ±0.23 | −0.4 | ±0.20 |
| P | NS | | NS | | NS | |
| EMPAP | 25.4 | 25.1 | 25.2 | 25.5 | 25.7 | 25.3 |
| (cm $H_2O$) | −0.2 | ±0.24 | +0.3 | ±0.71 | −0.5 | ±0.36 |
| P | NS | | NS | | NS | |
| LVP | 136 | 136 | 132 | 130 | 131 | 128 |
| (mmHg) | −0.7 | ±0.42 | −1.8 | ±0.75 | −3.3 | ±1.67 |
| P | NS | | NS | | NS | |
| LVEDP | 1.3 | 1.3 | 1.8 | 2.7 | 1.8 | 2.3 |
| (mmHg) | 0 | ±0 | +0.8 | ±0.54 | +0.5 | ±0.50 |
| P | NS | | NS | | NS | |
| dp/dt | 3750 | 3646 | 3375 | 3229 | 3458 | 3333 |
| (mmHg/sec) | −104 | ±81.8 | −146 | ±59.7 | −125 | ±79.1 |
| P | NS | | NS | | NS | |
| MAP | 122 | 121 | 117 | 117 | 119 | 117 |
| (mmHg) | −1.0 | ±0.82 | −0.7 | ±0.88 | −2.0 | ±0.81 |
| P | NS | | NS | | NS | |
| MPAF | 1483 | 1467 | 1425 | 1400 | 1425 | 1400 |
| (ml/min) | −16.7 | ±10.54 | −25 | ±11.2 | −25 | ±11.2 |
| P | NS | | NS | | NS | |
| HR | 175 | 176 | 175 | 174 | 182 | 184 |
| (beats/min) | +0.7 | ±1.02 | −0.8 | ±2.1 | +2 | ±1.9 |
| P | NS | | NS | | NS | |
| PVR | 1042 | 1050 | 1021 | 1051 | 1017 | 1043 |
| (dynes·sec/$cm^5$) | +8 | ±5.81 | +30 | ±39.2 | +26 | ±26.4 |

*Each group of numbers includes control mean, mean for effect, difference ± standard error, and the significance level for five dogs.

TABLE 4.2 (continued)

**Interaction of Subthreshold Concentrations of Isopropanol and Trichloroethylene***

| | Isopropanol subthreshold (0.63 ± 0.13%) | | Trichloroethylene subthreshold (0.01%) | | Isopropanol and trichloroethylene | |
|---|---|---|---|---|---|---|
| | C | E | C | E | C | E |
| P | NS | | NS | | NS | |
| SVR (dynes•sec/cm⁵) | 6526 +30 | 6556 ±72.9 | 6561 +19.8 | 6580 ±114.8 | 6641 −36 | 6606 ±96.0 |
| P | NS | | NS | | NS | |
| Stroke volume (ml) | 8.9 −0.1 | 8.7 ±0.06 | 8.3 −0.1 | 8.2 ±0.08 | 8.1 −0.2 | 7.9 ±0.08 |
| P | NS | | NS | | 0.05 | |
| Stroke work (g•m) | 15.1 −0.4 | 14.8 ±0.08 | 13.3 −0.2 | 13.1 ±0.22 | 13.6 −0.5 | 13.1 ±0.18 |
| P | 0.01 | | NS | | 0.05 | |

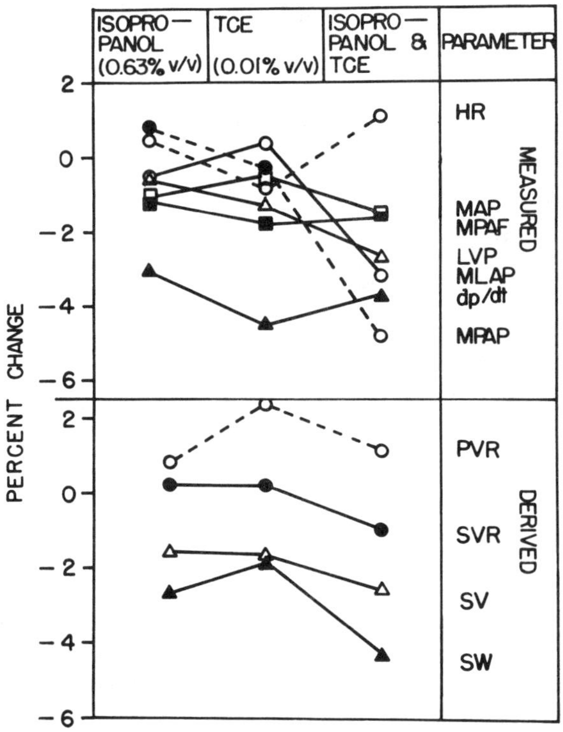


FIGURE 4.3. The effect of subthreshold concentration of isopropanol (0.63% v/v), subthreshold concentration of trichloroethylene (0.01% v/v), and a mixture of both, on various measured and derived parameters of the open-chest dog preparation.

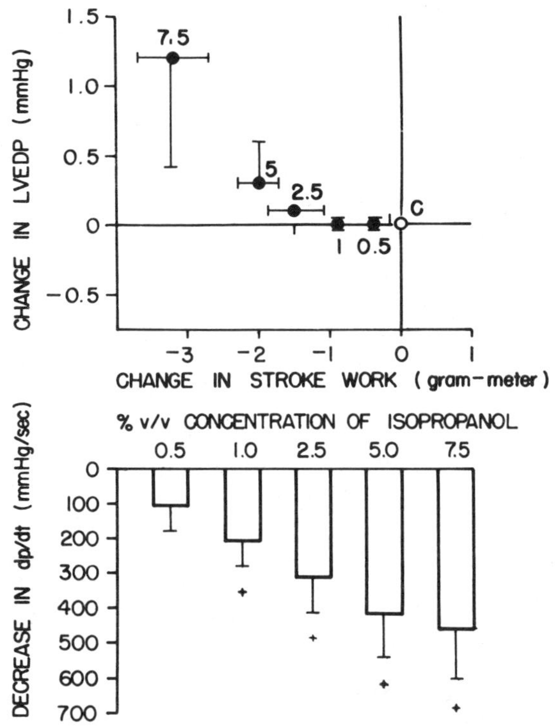


FIGURE 4.4. The effect of various concentrations of isopropanol on the myocardial function (above) and contractility (below). Note that significant decrease in left ventricular dp/dt was achieved by as low a concentration as 1% v/v of isopropanol.

Chapter 5

# SUMMARY OF CARDIAC OUTPUT AND SYSTEMIC VASCULAR EFFECTS OF ISOPROPANOL, FLUOROCARBONS, AND CHLORINATED SOLVENTS

The results of isopropanol in the canine preparation described in the preceding chapter relate to a third group of inhalants, specifically isopropanol. The two groups of inhalants previously investigated were three fluorocarbons[1 54] and two chlorinated solvents.[2] When the hemodynamic features of the six compounds belonging to three groups are compared in Table 5.1 and Figure 5.1, the following generalizations can be made:

**Threshold concentrations for reduction in cardiac output** – Methyl chloroform and trichloroethylene have the lowest inhaled concentration (0.1%) which causes a significant reduction of cardiac output measured by a flow probe inserted around the main pulmonary artery. A higher concentration (1.0%) is needed for fluorocarbon 11 (trichlorofluoromethane), and even more with the other inhalants: 2.5% for isopropanol and FC 114 (dichlorotetrafluoroethane) and 10% for FC 12 (dichlorodifluoromethane).

**Threshold concentrations for reduction in arterial blood pressure** – Four of the inhalants have similar threshold concentrations for a fall in both cardiac output and arterial blood pressure: 0.1% methyl chloroform, 1.0% FC 11, 2.5% FC 114, and 10% FC 12. A fifth inhalant has a small difference in threshold concentrations for both effects, i.e., trichloroethylene 0.1% reduces cardiac output but 0.25% also decreases arterial blood pressure. For isopropanol, there is a wide margin of threshold concentration for reduction in output and blood pressure; the respective concentrations are 2.5 and 7.5% for isopropanol. This alcohol is more likely to reduce cardiac output than produce hypotension.

**Calculated systemic vascular resistance** – Isopropanol is less likely to cause hypotension because the reduction in cardiac output is accompanied by an increase in systemic vascular resistance (vasoconstriction). On the other hand, the chlorinated solvents are more likely to produce hypotension since the reduction in cardiac output is usually accompanied by a decrease in systemic vascular resistance (vasodilation). The three fluorocarbons have varied effects on resistance, which is either increased or decreased depending on the concentration and the nature of the compound.

**Unknown nature of vascular effect** – The total systemic vascular resistance represents the sum of changes in caliber of the coronary, cerebral, renal, splanchnic, and limb blood vessels. The information on how these behave is incomplete. So far, vasodilation of the limb vessels by methyl chloroform and FC 11 has been demonstrated.[155] The effects of the other compounds on the same and neighboring vascular beds are unknown. Additional experiments are needed to complete the hemodynamic profile for these solvents especially because hypotension has been reported in poisoning from solvents in general, and from isopropanol in particular.

TABLE 5.1

**Summary of Effects of Solvents and Propellants on Cardiac Output, Systemic Arterial Blood Pressure, and Systemic Vascular Resistance (mean changes from control)**

| | Alcohol | Fluorocarbons | | | Chlorinated solvents | |
|---|---|---|---|---|---|---|
| Concentration | Isopropanol | FC 11 | FC 12 | FC 114 | Methyl chloroform | Trichloroethylene |
| 0.01% | | | | | | 0 ml/min<br>−1 mm Hg<br>−308 dynes·sec/cm$^5$ |
| 0.05% | | | | | −17 ml/min<br>−1 mm Hg<br>−69 dynes·sec/cm$^5$ | −33 ml/min<br>−3 mm Hg<br>−220 dynes·sec/cm$^5$ |
| 0.10% | | | | | −55 ml/min*<br>−3 mm Hg*<br>+112 dynes·sec/cm$^5$ | −67 ml/min*<br>−6 mm Hg<br>−330 dynes·sec/cm$^5$ |
| 0.25% | | | | | −79 ml/min<br>−7 mm Hg<br>+28 dynes·sec/cm$^5$ | −125 ml/min<br>−8 mm Hg*<br>−72 dynes·sec/cm$^5$ |
| 0.50% | −25 ml/min<br>0 mm Hg<br>+117 dynes·sec/cm$^5$ | −24 ml/min<br>−1 mm Hg<br>+44 dynes·sec/cm$^5$ | | | −129 ml/min<br>−17 mm Hg<br>−216 dynes·sec/cm$^5$ | −133 ml/min<br>−9.0 mm Hg<br>−128 dynes·sec/cm$^5$ |
| 1.0% | −50 ml/min<br>−1.3 mm Hg<br>+170 dynes·sec/cm$^5$ | −52 ml/min*<br>−4 mm Hg*<br>+30 dynes·sec/cm$^5$ | | −17 ml/min<br>0 mm Hg<br>−64 dynes·sec/cm$^5$ | −171 ml/min<br>−29 mm Hg<br>−875 dynes·sec/cm$^5$ * | −275 ml/min<br>−19 mm Hg<br>−571 dynes·sec/cm$^5$ |
| 2.5% | −84 ml/min*<br>−2.2 mm Hg<br>+288 dynes·sec/cm$^5$ | −129 ml/min<br>−8 mm Hg<br>+336 dynes·sec/cm$^5$ | | −92 ml/min*<br>−6 mm Hg*<br>+207 dynes·sec/cm$^5$ | | |
| 5.0% | −100 ml/min<br>−2.7 mm Hg<br>+272 dynes·sec/cm$^5$ | −248 ml/min<br>−24 mm Hg<br>−167 dynes·sec/cm$^5$ | −24 ml/min<br>−3 mm Hg<br>−59 dynes·sec/cm$^5$ | −100 ml/min<br>−9 mm Hg<br>+253 dynes·sec/cm$^5$ | | |
| 7.5% | −150 ml/min<br>−5.2 mm Hg*<br>+387 dynes·sec/cm$^5$ | | | | | |
| 10.0% | | | −104 ml/min*<br>−8 mm Hg*<br>*187 dynes·sec/cm$^5$ | | | |
| 20.0% | | | −151 ml/min<br>−13 mm Hg<br>+393 dynes·sec/cm$^5$ | −183 ml/min<br>−24 mm Hg<br>+137 dynes·sec/cm$^5$ | | |

$*P < 0.05$

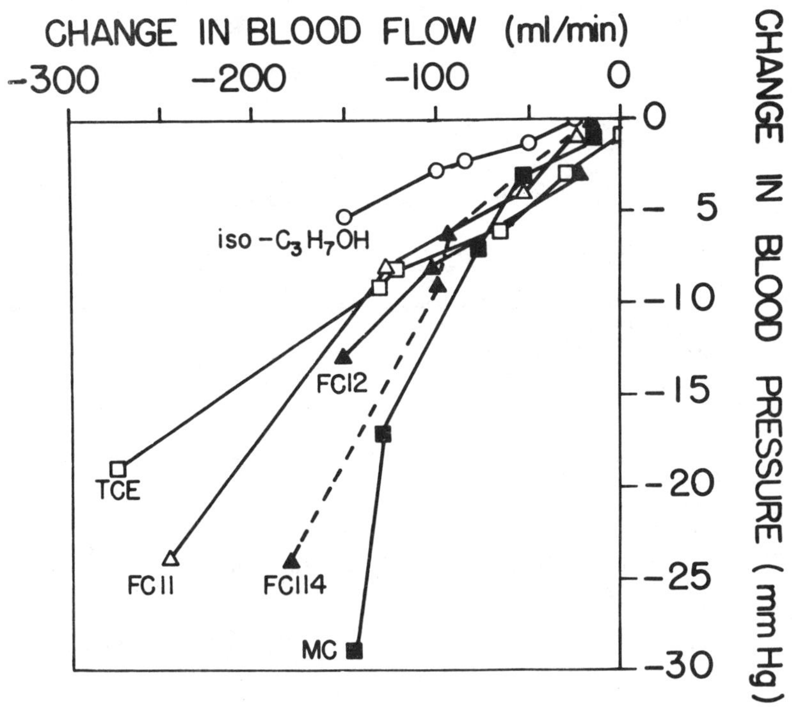


FIGURE 5.1. The changes in pulmonary blood flow (ml/min) are plotted versus changes in blood pressure (mm Hg), under the influence of isopropanol(iso-$C_3H_7OH$), halogenated solvents (methyl chloroform MC, trichloroethylene TCE), and the fluorinated propellants (FC 11, FC 12 and FC 114).

# REFERENCES – Isopropanol


1. **Aviado, D. M.,** Toxicity of propellants, in *Progress in Drug Research,* Vol. 18, Jucker, E. Ed., Birkhäuser Verlag, Basel, 1974, 365.
2. **Aviado, D. M., Zakhari, S., Simaan, J. A., and Ulsamer, A. G.,** *Methyl Chloroform and Trichloroethylene in the Environment,* Golberg, L., Ed., CRC Press, Cleveland, 1976, p. 17–36, 61–73.
3. **Hatch, L. F.,** *Isopropyl Alcohol,* McGraw-Hill, New York, 1961.
4. **Browning, E.,** *Toxicity of Industrial Organic Solvents,* Her Majesty's Stationery Office, London, revised edition, 1952.
5. **Browning, E.,** *Toxicity and Metabolism of Industrial Solvents,* Elsevier, Amsterdam, 1965.
6. **Baker, R. N., Alenty, A. L., and Zack, J. F.,** Toxic volatiles in alcoholic coma, a report of simultaneous determination of blood methanol, ethanol, isopropanol, acetaldehyde and acetone by gas chromatography, *Bull. Los Angeles Neurol. Soc.,* 33,140,1968.
7. **Hessel, D. W. and Modglin, F. R.,** The quantitative determination of ethanol and other volatile substances in blood by gas-liquid partition chromatography, *J. Forensic Sci.,* 9, 255, 1964.
8. **McCord, W. M. and Gadsden, R. H.,** Identification and determination of alcohols in blood by gas chromatography, *J. Gas Chromatogr.,* 2, 38, 1964.
9. **Mather, A. and Assimos, A.,** Evaluation of gas-liquid chromatography in assays for blood volatiles, *Clin. Chem.* (N.Y.), 11, 1023, 1965.
10. **Freudiger, J. B. and Vignau, J. A.,** Determination of alcohols in body fluids by gas-liquid chromatography, *J. Forensic Sci.,* 10, 73, 1965.
11. **Natelson, S. and Stellate, R. L.,** Instrumentation for the concentration of trace components of a mixture for gas chromatography; application to the determination of acetone, ethanol, methanol, and 2-propanol in blood, *Microchem. J.,* 9, 245, 1965.
12. **Schenkel, S. and Schwartz, H.,** A note on the detection of isopropyl alcohol in tissue, *Am J. Clin. Pathol.,* 23, 716, 1953.
13. **Falcone, N.,** Gas chromatographic determination of ethanol and isopropanol or acetone in drugs, *J. Assoc. Off. Anal. Chem.,* 56, 684, 1973.
14. **John, Sister M.,** Isopropyl alcohol's usefulness in hospitals analyzed, *Hosp. Manage.,* 57, 86, 1944.
15. **Straughn, W. R.,** Isopropyl alcohol as a disinfectant, *Mod. Hosp.,* 66, 90, 1946.
16. **Tanner, F. W. and Wilson, F. L.,** Germicidal action of aliphatic alcohols, *Proc. Soc. Exp. Biol. Med.,* 52, 138, 1943.
17. **Wantia, J.,** Über die Desinfektions- und Reinigungswirkung des Methyl- Äthyl- und Isopropylalkohols in Benzin, *Arch. Hyg.,* 130, 129, 1943.
18. **Powell, H. M.,** The antiseptic properties of isopropyl alcohol in relation to cold sterilization, *J. Indiana State Med. Assoc.,* 38, 303, 1945.
19. **Nancekievill, D. G.,** Decontamination of the Cape ventilator, *Anaesthesia,* 24, 359, 1969.
20. **Jones, E. H.,** Prevention of "swimming pool ear," *Laryngoscope,* 81, 731, 1971.
21. **Meyer, F. and Klink, F.,** Die Pharmakologie des n-Propylalkohols und seine Verwendungsmoglichkeit in pharmazeutischen Zubereitungen, *Arzneim. Forsch.,* 5, 513, 1955.
22. **Barman, M. L., Sigel, N. B., Beedle, D. B. and Larson, R. K.,** Acute and chronic effects of glue sniffing, *Calif. Med.,* 100, 19, 1964.
23. **Ruemele, T.,** Masking the odour of cosmetic ingredients, *Perfum. Essent. Oil Rec.,* 41, 43, 1950.
24. **Lechnitz, V.,** Science and practice in cosmetics, *Kosmet. Aerosole,* 44, 65, 1971.
25. **McAllister, W. G.,** Stain removers and household dry cleaning fluids, *Soap, Perfum. Cosmet.,* 22, 1115, 1949.
26. **Hackett, W. J.,** Carpet shampoos, *Manuf. Chem. Aerosol News,* 42, 50, 1971.
27. **Schlueter, W.,** Spot and stain removing compositions, German Patent 2,042,482, 1972.
28. **Henderson, R. M. and Habermann, W. F.,** Concentrating fish protein, German Patent 2,014,214, 1971.
29. **Gleason, M. N., Gosselin, R. E., Hodge, H. C., and Smith, R. P.,** *Clinical Toxicology of Commercial Products,* Williams & Wilkins, Baltimore, 1969.
30. **Schwarzkopf, F. H.,** Preparation of a hair-conditioner containing an alcohol, Netherlands Appl. 6,607,794, 1966.
31. **Hsiung, D. Y.,** Dispenser for hair-treating compositions, (Gillette Co.). German Patent 2,239,690, 1973; U.S. Appl 171,398, 1971.
32. **Takamine, K. and Shimizu, H.,** Alcohol-water hair spray aerosol, (Osaka Aerosol Industry Co., Ltd.). Japanese Patent 70 11, 213, 1970.
33. **Strand, D. L. and Abler, R. L.,** Heavy-duty foaming aerosol abrasive cleaner compositions, (Minnesota Mining and Manufacturing Co.) U.S. Patent 3,650,956, 1972.
34. **Kessler, A., Layne, G. L., and Spector, C. L.,** Shoe polishes, (Proctor and Gamble Co.). U.S. Patent 3,231,397, 1966.
35. **Kaufman, K. L., Martin, D. N., and Brown, W. J.,** Aerosol spray for dewrinkling fabrics, (CPC International Inc.). U.S. Patent 3,600,325, 1971.
36. **Tsetlin, V. M., Bessanova, I. V., and Zhuk, E. B.,** Selection of solvents for insecticide aerosol cans, *Khim. Promst.* (Moscow), 47, 31, 1971.

37. **Woogerd, S. M.,** Surface active and buffering adjuvants for agricultural sprays, (Hercules Glue Co., Ltd.). U.S. Patent 3,244,502, 1966.
38. **Augart, H.,** Stable suspension aerosols for pharmaceutical sprays, (Goedecke A.G.) German Patent 1,935,939, 1971.
39. **Kulkarni, A. B., Carrigan, P. J., Ward, J. B., and Hutchins, H. H.,** Quantitative evaluation of chilling effect of aerosol sprays, *J. Pharm. Sci.,* 58, 481, 1969.
40. **Adelson, L.,** Fatal intoxication with isopropyl alcohol (rubbing alcohol), *Am. J. Clin. Pathol.,* 38, 144, 1962.
41. **McCord, W. M., Switzer, P. K., and Brill, H. H.,** Isopropyl alcohol intoxication, *South. Med. J.,* 41, 639, 1948.
42. **Chapin, M. A.,** Iospropyl alcohol poisoning with acute renal insufficiency, *J. Maine Med. Assoc.,* 40, 288, 1949.
43. **Juncos, L. and Taguchi, J. T.,** Isopropyl alcohol intoxication, report of a case associated with myopathy renal failure, and hemolytic anemia, *JAMA,* 203, 732, 1968.
44. **Wills, J. H., Jameson, E. M., and Coulston, F.,** Effects on man of daily ingestion of small doses of isopropyl alcohol, *Toxicol. Appl. Pharmacol.,* 15, 560, 1969.
45. **Kemal, H.,** Beitrag zur Kenntnis der Schicksale des Isopropylalkohols im menschliehen Organismus, *Biochem. Z.,* 187, 461, 1927.
46. **Corbett, J. and Meier, G.,** Suicide attempted by rectal administration of drug, *JAMA,* 206, 2320, 1968.
47. **Newman, A. J.,** Isopropanol administered rectally, *A.M.A. Am. J. Dis. Child.,* 81, 798, 1951.
48. **McFadden, S. W. and Haddow, J. E.,** Coma produced by topical application of isopropanol, *Pediatrics,* 43, 622, 1969.
49. **Wise, J. R., Jr.,** Alcohol sponge baths, *N. Engl. J. Med.,* 280, 840, 1969.
50. **Moss, M. H.,** Alcohol-induced hypoglycemia and coma caused by alcohol sponging, *Pediatrics,* 46, 445, 1970.
51. **Garrison, R. F.,** Acute poisoning from the use of isopropyl alcohol in tepid sponging, *JAMA,* 152, 317, 1953.
52. **Vermeulen, R.,** Isopropyl alcohol and diabetes, *Pa. Med.,* 69, 53, 1966.
53. **Wax, J., Ellis, F. W., and Lehman, A. J.,** Absorption and distribution of isopropyl alcohol, *J. Pharmacol. Exp. Ther.,* 97, 229, 1949.
54. **Lehman, A. J., Schwerma, H., and Rickards, E.,** Isopropyl alcohol: rate of disappearance from the blood stream of dogs after intravenous and oral administration *J. Pharmacol. Exp. Ther.,* 82, 196, 1944.
55. **Orskov, S. L.,** Experiments on the oxidation of propyl alcohol in rabbits, *Acta Physiol. Scand.,* 20, 258, 1950.
56. **Berggren, S. M.,** Uber die Umzetzung des *n*-Propyl und *n*-Butyl-alkoholes im Organismus, nebst Bemerkungen uber die Oxydationsprodukte der Alkohole bei oxydation im Biochromat-Schwefelsauregemisch, *Skand. Arch. Physiol.,* 78, 249, 1938.
57. **Lehman, A. J., Schwerma, H., and Rickards, E. E.,** Isopropyl alcohol, acquired tolerance in dogs, rate of disappearance from blood stream in various species, and effects on successive generations of rats, *J. Pharmacol. Exp. Ther.,* 85, 61, 1945.
58. **Nordmann, R., Ribiere, C., Rouach, H., Beauge, F., Giudicelli, Y., and Nordmann, J.,** Metabolic pathways involved in the oxidation of isopropanol into acetone by the intact rat, *Life Sci.,* 13, 919, 1973.
59. **Lehman, A. J.,** Effect of insulin, insulin-dextrose, and water diuresis on metabolism of isopropyl alcohol, *Proc. Soc, Exp. Biol. Med.,* 62, 232, 1946.
60. **Lester, D. and Benson, G. D.,** Alcohol oxidation in rats inhibited by pyrazole, oximes, and amides, *Science,* 169, 282, 1970.
61. **Dalziel, K. and Dickinson, F. M.,** The kinetics and mechanism of liver alcohol dehydrogenase with primary and secondary alcohols as substrates, *Biochem. J.,* 100, 34, 1966.
62. **Abshagen, U. and Riegbrock, N.,** Elimination von 2-Propanol und ihre Beeinflussung durch aliphatische alkohole, *Naunyn Schmiedeberg's Arch. Pharmacol.,* 264, 212, 1969.
63. **Abshagen, U. and Rietbrock, N.,** Zum Mechanismus der 2-Propanoloxydation, Interferenzversuche mit niederen aliphatischen Alkoholen in vivo und an der isoliert perfundierten Rattenleber, *Naunyn Schmiedeberg's Arch. Pharmacol.,* 265, 411, 1970.
64. **Conover, T.,** Influence of organic solutes on the reactions of oxidative phosphorylation, *J. Biol. Chem.,* 244, 254, 1969.
65. **Ashkar, F. S. and Miller, R.,** Hospital ketosis in the alcoholic diabetic, a syndrome due to IP alcohol intoxication, *South. Med. J.,* 64, 1409, 1971.
66. **Teschke, R., Hasumura, Y., and Lieber, C. S.,** NADPH-dependent oxidation of methanol, ethanol, propanol and butanol by hepatic microsomes, *Biochem. Biophys. Res. Commun.,* 60, 851, 1974.
67. **Gaillard, D. and Derache, R.,** Vitesse de la metabolisation de differents alcools chez le rat, *C. R. Soc. Biol.* (Paris), 158, 1605, 1964.
68. **Morris, H. J. and Lightbody, H. D.,** The toxicity of isopropanol, *J. Ind. Hyg. Toxicol.,* 20, 428, 1938.
69. **Lehman, A. J. and Chase, H. F.,** The acute and chronic toxicity of isopropyl alcohol, *J. Lab. Clin. Med.,* 29, 561, 1944.
70. **Macht, D. I.,** Pharmacological examination of isopropyl alcohol, *Arch. Int. Pharmacodyn. Ther.,* 26, 285, 1922.
71. **Grant, D. H.,** The pharmacology of isopropyl alcohol, *J. Lab. Clin. Med.,* 8, 382, 1923.
72. **Boruttau, H.,** Die Verwendung von Isopropylalkohol zu hygienischen und kosmetischen Zweeken, *Deutsch. Med. Wochenschr.,* 47, 747, 1921.

73. **Gerarde, H. W. and Ahlstrom, D. B.**, The aspiration hazard and toxicity of a homologous series of alcohols, *Arch. Environ. Health,* 13, 457, 1966.
74. **Kimura, E. T., Ebert, D. M., and Dodge, P. W.**, Acute toxicity and limits of solvent residue for sixteen organic solvents, *Toxicol. Appl. Pharmacol.,* 19, 699, 1971.
75. **Smyth, H. F. and Carpenter, C. P.**, Further experience with the range finding test in the industrial toxicology laboratory, *J. Ind. Hyg. Toxicol.,* 30, 63, 1948.
76. **Boughton, L. L.**, The relative toxicity of ethyl and isopropyl alcohols as determined by long term rat feeding and external application, *J. Am. Pharm. Assoc.,* 33, 111, 1944.
77. **Hufferd, R. W.**, Variation in physiological activity of alcohols among isomers and homologs, *J. Am. Pharm. Assoc.,* 21, 549, 1932.
78. **Paget, M. and Mestre, A.**, Sur la toxicite de l'alcool isopropylique, *Bull. Sci. Pharmacol.,* 49, 17, 1942.
79. **Wallgren, H.**, Relative intoxicating effects on rats of ethyl, propyl, and butyl alcohols, *Acta Pharmacol. Toxicol.,* 16, 217, 1960.
80. **MacGregor, D. C., Armour, J. A., Goldman, B. S., and Bigelow, W. G.**, The effects of ether, ethanol, propanol, and butanol on tolerance to deep hypothermia, *Dis. Chest,* 50, 523, 1966.
81. **Weese, H.**, Comparative study of the activity and toxicity of the vapors of the lower aliphatic alcohols, *Arch. Exp. Pathol. Pharmakol.,* 135, 118, 1928.
82. **Macht, D. I.**, Isopropyl alcohol, a convenient laboratory anesthetic for cats, *Proc. Soc. Exp. Biol. Med.,* 19, 85, 1921.
83. **Seeman, P.**, Erythrocyte membrane stabilization by steroids and alcohols; a possible model for anesthesia, *Biochem. Pharmacol.,* 15, 1632, 1966.
84. **Glasgow, J. F. T. and Ferris, J. A. J.**, Encephalopathy and visceral fatty infiltration of probable toxic aetiology, *Lancet,* 1, 451, 1968.
85. **Fuller, H. C. and Hunter, O. B.**, Isopropyl alcohol – an investigation of its physiologic properties, *J. Lab. Clin. Med.,* 12, 326, 1927.
86. **Lindbohm, R. and Wallgren, H.**, Changes in respiration of rat brain cortex slices induced by some aliphatic alcohols, *Acta Pharmacol. Toxicol.,* 19, 53, 1962.
87. **Majchrowicz, E.**, Effects of aliphatic alcohols and aldehydes on the metabolism of potassium-stimulated rat brain cortex slices, *Can. J. Biochem.,* 43, 1041, 1965.
88. **Israel, Y., Kalant, H., and LeBlanc, A. E.**, Effects of lower alcohols on potassium transport and microsomal adenosine-triphosphatase activity of rat cerebral cortex, *Biochem. J.,* 100, 27, 1966.
89. **Armstrong, C. M. and Binstock, L.**, The effects of several alcohols on the properties of the squid giant axon, *J. Gen. Physiol.,* 48, 265, 1964.
90. **Skou, J. C.**, Relation between the ability of various compounds to block nervous conduction and their penetration into a monomolecular layer of nerve tissue lipoids, *Biochem. Biophys. Acta,* 30, 625, 1958.
91. **Rumsby, M. B. and Finean, J. B.**, The action of organic solvents on the myelin sheath of peripheral nerve tissue. II. Short-chain aliphatic alcohols, *J. Neurochem.,* 13, 1509, 1966.
92. **Gage, F. W.**, The effect of methyl ethyl and n-propyl alcohol on neuromuscular transmission in the rat, *J. Pharmacol. Exp. Ther.,* 150, 236, 1965.
93. **Visudhiphan, P. and Kaufman, H.**, Increased cerebrospinal fluid protein following isopropyl alcohol intoxication, *N.Y. State J. Med.,* 71, 887, 1971.
94. **Schaffarzick, R. W.**, Anticonvulsant action of isopropyl alcohol, *Proc. Soc. Exp. Biol. Med.,* 74, 211, 1950.
95. **Chu, N., Driver, R. L., and Hanzlik, P. J.**, Anticonvulsant action of isopropyl alcohol, *J. Pharmacol. Exp. Ther.,* 92, 291, 1948.
96. **Driver, R. L.**, Isopropyl alcohol, other ketogens, and miscellaneous agents on thresholds for electrical convulsions and diphenylhydantoin, *Proc. Soc. Exp. Biol. Med.,* 64, 248, 1947.
97. **Marcus, J.**, Stereochemical theory of smell, *Yale Sci.,* 44, 7, 1970.
98. **Stone, H., Carregal, E. J. A., and Williams, B.**, The olfactory-trigeminal response to odorants, *Life Sci.,* 5, 2195, 1966.
99. **Nakano, J. and Moore, S. E.**, Effect of different alcohols on the contractile force of the isolated guinea-pig myocardium, *Eur. J. Pharmacol.,* 20, 266, 1972.
100. **Ringer, S. and Sainsbury, H.**, Observations on the relative effects of certain members of the ethylic alcohol series on the ventricle of the frog's heart, *Practitioner,* 30, 339, 1883.
101. **Steele, R. H. and Wilhelm, D. L.**, The inflammatory reaction in chemical injury. I. Increased vascular permeability and erythema induced by various chemicals, *Br. J. Exp. Pathol.,* 47, 612, 1966.
102. **Gerarde, H. W. and Eckardt, R. E.**, Aspiration hazards of petroleum products in children, *14th Int. Congr. Occupational Health,* 2, 723, 1964.
103. **Nordmann, R., Giudicelli, Y., Beaugé, F., Clement, M., Ribiere, C., Rouach, H., and Nordmann, J.**, Studies on the mechanisms involved in the isopropanol-induced fatty liver, *Biochem. Biophys. Acta,* 326, 1, 1973.
104. **Ammon, H. P. T., Heim, F., Estler, C. -J., Fickeis, G., and Wagner, M.**, The influence of aliphatic alcohols and their halogen derivatives on the coenzyme A in the liver of mice, *Biochem. Pharmacol.,* 16, 1533, 1967.

105. **Giudicelli, Y., Nordmann, R., and Nordmann, J.,** Modifications of plasma post-heparine lipolytic activity and tissue lipoprotein lipase activity induced in the rat by acute administration of ethanol or propan-2-ol, *Clin. Sci. Mol. Med.,* 48, 153, 1975.
106. **Mallov, S. and Bloch, J. L.,** Role of hypophysis and adrenals in fatty infiltration of liver resulting from acute ethanol intoxication, *Am. J. Physiol.,* 184, 29, 1956.
107. **Beaugé, F., Clement, M., Giudicelli, Y., Nordmann, R., and Nordmann, J.,** Effet de l'administration d'isopropanol sur l'incorporation du palmitate 1-$^{14}$C dans les triglycerides et phospholipides hepatiques chez le rat, *C. R. Acad. Sci.* (Paris), 275, 3005, 1972.
108. **Beaugé, F., Clement, M., Nordmann, J., and Nordmann, R.,** Perturbations du metabolisme hepatique du palmitate [1-$^{14}$C] determinees par l'administration de n-propanol chez le rat, *Biochimie,* 56, 1157, 1974.
109. **Gaillard, D. and Derache, R.,** Action de quelques alcools aliphatiques sur la mobilisation de differentes fractions lipidiques chez la rate, *Food Cosmet. Toxicol.,* 4, 515, 1966.
110. **Gaillard, D. and Derache, R.,** Effets d'une intoxication aigue, par differents alcools, sur les fractions lipidiques du foie, chez la ratte, *C. R. Acad. Sci.* (Paris), 261, 3880, 1965.
111. **Plaa, G. L. and Traiger, G. J.,** Mechanism of potentiation of $CCl_4$-induced hepatotoxicity, *Int. Congr. Pharmacol.,* 2, 100, 1972.
112. **Traiger, G. J. and Plaa, G. L.,** Effect of aminotriazole on isopropanol- and acetone-induced potentiation of $CCl_4$ hepatotoxicity, *Can. J. Physiol. Pharmacol.,* 51, 291, 1973.
113. **Traiger, G. J. and Plaa, G. L.,** Chlorinated hydrocarbon toxicity, potentiation by isopropyl alcohol and acetone, *Arch. Environ. Health,* 28, 276, 1974.
114. **Traiger, G. J. and Plaa, G. L.,** Differences in the potentiation of carbon tetrachloride in rats by ethanol and isopropanol pretreatment, *Toxicol. Appl. Pharmacol.,* 20, 105, 1971.
115. **Traiger, G. J. and Plaa, G. L.,** Effect of isopropanol on $CCl_4$-induced changes in perfused rat liver hemodynamics, *Arch. Int. Pharmacodyn. Ther.,* 202, 102, 1973.
116. **Traiger, G. J. and Plaa, G. L.,** Relationship of alcohol metabolism to the potentiation of $CCl_4$ hepatotoxicity induced by aliphatic alcohols, *J. Pharmacol. Exp. Ther.,* 183, 481, 1972.
117. **Cornish, H. H. and Adefuin, J.,** Potentiation of carbon tetrachloride toxicity by aliphatic alcohols, *Arch. Environ. Health,* 14, 447, 1967.
118. **Sipes, I. G., Stripp, B., Krishna, G., Maling, H. M., and Gillette, J. R.,** Enhanced hepatic microsomal activity by pretreatment of rats with acetone or isopropanol, *Proc. Soc. Exp. Biol. Med.,* 142, 237, 1973.
119. **De Breyer, I. J. J. and Soehring, K.,** Wirkungen verschiedener Alkohole auf die oxidative desaminierung durch Leberschnitte, *Arzneim. Forsch.,* 18, 740, 1968.
120. **Ortiz, A., Soehring, K., and Teske, E.,** The effects of ethanol, n-butanol and isopropanol upon 5-hydroxytryptamine metabolism in the perfused frog liver, *Arzneim. Forsch.,* 21, 118, 1971.
121. **Ortiz, A. and Soehring, K.,** The effects of alcohols upon oxidative deamination and conjugation of 5-hydroxy-tryptamine, *Arzneim. Forsch.,* 21, 120, 1971.
122. **Shore, J. D. and Theorell, H.,** Substrate inhibition in the liver alcohol dehydrogenase reaction, *Arch. Biochem.,* 117, 375, 1966.
123. **Forsander, O. A.,** Influence of some aliphatic alcohols on the metabolism of rat liver slices, *Biochem. J.,* 105, 93, 1967.
124. **Ho, Y. H., Schwarze, I., and Soehring, K.,** Effect of low molecular weight aliphatic alcohols on chloral hydrate metabolism in rat liver slices, *Arzneim. Forsch.,* 20, 1507, 1970.
125. **Koivusalo, M.,** Influence of other alcohols on the oxidation of methanol to formaldehyde in rat liver homogenates, *Ann. Med. Exp. Biol. Fenn.,* 41, 251, 1963.
126. **Gotterer, G. S.,** Rat liver D-β-hydroxybutyrate dehydrogenase. III. Inhibition by topical anesthetics, *Biochemistry,* 8, 641, 1969.
127. **Oura, E., Raiha, N. C. R., and Suomalainen, H.,** Influence of some alcohols and narcotics on the adenosine phosphates in the liver of the mouse, *Ann. Med. Exp. Biol. Fenn.,* 45, 57, 1967.
128. **Levy-Hochman, S.,** Kidney bleeding produced by alcohols and some anesthetics, *Exp. Med. Surg.,* 2, 216, 1944.
129. **Leng, C. O.,** A method for evaluating the diuretic response of alcohols and related substances in the rat, *Far East Med. J.,* 6, 141, 1970.
130. **Baer, P. N. and Archard, H. O.,** Factitial disease of the gingiva and buccal mucosa, *N.Y. State Dent. J.,* 40, 33, 1974.
131. **Weisbrodt, N. W., Kienzle, M., and Cooke, A. R.,** Comparative effects of aliphatic alcohols on the gastric mucosa, *Proc. Soc. Exp. Biol. Med.,* 142, 450, 1973.
132. **Tang, J.,** Competitive inhibition of pepsin by aliphatic alcohols, *J. Biol. Chem.,* 240, 3810, 1965.
133. **Gibel, W., Lohs, K. H., Wildner, G. P., and Schramm, T.,** Experimentelle Untersuchungen zur kanzerogenen Wirkung hoherer Alkohole am Beispiel von 3-Methyl-1-butanol, 1-Propanol und 2-Methyl-1-propanol, *Z. Exp. Chir.,* 7, 235, 1974.
134. **Baader, E. W.,** Die Gefahr des Berufskrebses bei der Isopropylalkoholfabrikation, *Zentralbl. Arbeitsmed. Arbeitsschutz,* 1, 90, 1951.
135. **Weil, C. S., Smyth, H. F., and Nale, T. W.,** Quest for a suspected industrial carcinogen, *A.M.A. Arch. Indust, Hyg. Occup. Med.,* 5, 535, 1952.
136. **Bittersohl, G.,** Kanzerogene Wirkung von "Isopropylol," *Arch. Geschwulstforsch.,* 43, 250, 1974.

137. **Fregert, S., Groth, O., Gruvberger, B., Magnusson, B., Mobacken, H., Rosrman, H.,** Hypersensitivity to secondary alcohols, *Acta Derm. Venereol.*, 51, 271, 1971.
138. **McInnes, A.,** Skin reaction to isopropyl alcohol, *Br. Med. J.*, 1, 357, 1973.
139. **Blank, I. H.,** Penetration of low-molecular-weight alcohols into skin. I. Effect of concentration of alcohol and type of vehicle, *J. Invest. Dermatol.*, 43, 415, 1964.
140. **Suzuki, M. and Arai, H.,** Relation between the chemical structures of alcohols and the increased capillary permeabilities on skin, *Nippon Yakurigaku Zasshi,* 62, 241, 1966.
141. **Poyser, R. H. and West, G. B.,** Structural requirements of sugars as antagonists of the vascular response to dextran in rat skin, *Br. J. Pharmacol. Chemother.*, 32, 219, 1968.
142. **Von Oettengen, W. F.,** Poisoning – a guide to clinical diagnosis and treatment, in *Human Investigation,* W. B. Saunders, Philadelphia, 1958, 393.
143. **Keeser, E.,** Uber den Isopropylalkohol und seine gewerbetoxikologische Beurteilung, *Zentralbl. Arbeitsmed. Arbeitsschutz,* 1, 25, 1951.
144. **Gardner, A. W.,** Aliphatic solvents, *Ann. Occup. Hyg.*, 9, 209, 1966.
145. **Freireich, A. W., Cinque, T. J., Xanthaky, G., and Landau, D.,** Hemodialysis for isopropanol poisoning, *New Engl. J. Med.*, 277, 699, 1967.
146. **King, L. H., Bradley, K. P., Shires, D. L., Jr.,** Hemodialysis for isopropyl alcohol poisoning, *JAMA,* 211, 1855, 1970.
147. **Dua, S. L.,** Peritoneal dialysis for isopropyl alcohol poisoning, *JAMA,* 230, 35, 1974.
148. American Conference of Governmental Industrial Hygienists, Documentation of the Threshold Limit Values for Substances in Workroom Air, 3rd ed., Cincinnati, Ohio, 1971, 141.
149. **Nelson, K. W., Ege, J. F., Jr., Ross, M., Woodman, L. E., and Silverman, L.,** Sensory response to certain industrial solvent vapors, *J. Ind. Hyg. and Toxicol.*, 25, 282, 1943.
150. **Baikov, B. K., Gorlova, O. E., Gusev, M. I., Novikov, Y. V., Yudina, T. V., and Sergeev, A. N.,** Hygienic standardization of the average daily maximum permissible concentrations of propyl and isopropyl alcohols in the atmosphere, *Gig. Sanit.*, 4, 6, 1974.
151. **Gorlova, O. E.,** Hygienic assessment of isopropyl alcohol as an atmospheric pollutant, *Gig. Sanit.*, 35, 9, 1970.
152. **Weese, H.,** Vergleichende Untersuchungen uber die Wirksamkeit und Giftigkeit der Dampfe niederer aliphatischer Alkohole, *Arch. Exp. Pathol. Pharmakol.*, 35, 118, 1928.
153. **Finney, D. J.,** *Probit Analysis,* 2nd ed., Cambridge University Press, London, 1964.
154. **Simaan, J. A. and Aviado, D. M.,** Hemodynamic effects of aerosol propellants. I. Cardiac depression in the dog, *Toxicology,* 5, 127, 1975.
155. **Simaan, J. A. and Aviado, D. M.,** Hemodynamic effects of aerosol propellants. III. Vascular resistance in the canine hind limb, *Toxicology,* 5, 287, 1976.

# *Part 2*
# *Methyl Ethyl Ketone*

## Chapter 6
# REVIEW OF THE LITERATURE ON METHYL ETHYL KETONE

Methyl ethyl ketone (MEK) is a highly volatile solvent, used in a variety of industrial processes and consumer products. Some applications listed in 1938 by Langedijk[1] and in 1953 by Browning[2] included the manufacture of lacquers, synthetic colorless resins, artificial leather, and lubricating oils. More recently, its usage has been significantly increased, particularly in polymerizing processes during plastics manufacture. Additionally, MEK has been identified as a commonly occurring carbonyl compound in an extensive range of basic foodstuffs.

Historically, MEK was considered particularly useful because of its ability to substitute for toxic solvents, such as benzol.[3] While MEK is believed relatively safe, extensive usage, especially in combination with acetone in the manufacture of synthetic ("American-made") rubber during and following World War II, led to accounts of health hazards attributable to MEK. It became apparent that workers could develop dermatitis from direct contact with the skin.[3] The vapor, besides having an offensive odor, irritates the eyes and nose,[5-7] and can itself produce dermatitis.[3] High vapor concentrations, encountered by accidental exposure or deliberate inhalation (as in "glue sniffing"), caused narcosis, and even unconsciousness or convulsions in some extreme cases.[3,4,8,9] It was also suspected that MEK might depress respiration, blood pressure, and body temperature.[10]

Although considered only moderately toxic as compared to the highly toxic solvents for which it has been substituted, (such as benzol[3]), the widespread human contact with MEK vapors during industrial and home use, as well as by intentional inhalation ("glue sniffing"),[4] warrants further examination of its toxicologic properties. To fulfill this need, studies of MEK by inhalational, intraperitoneal, and oral routes have been carried out in mice (Chapter 7). The current monograph also reports on a series of experiments in the dog and cat designed to investigate effects of MEK on the heart, lungs, and circulation (Chapters 8 and 9).

The physical, chemical, and biological properties of MEK have been investigated since the late 19th century. A review of the literature is presented to summarize these studies, with emphasis on the toxicities of MEK thus far elucidated. Topics are discussed under the following headings: (A) Chemical and physical properties; (B) Medical uses; (C) Industrial uses; (D) Consumer products; (E) Human investigation; (F) Metabolism and disposition in animals; (G) Toxicologic investigation; (H) Pharmacodynamics; (I) Clinical toxicology; (J) Threshold limit values; and (K) Conversion table. Where possible, comparison of biological effects is made with acetone and other common organic solvents. Concentrations of MEK given in percent (v/v) or ppm in the literature have also been expressed in mg/l (w/v) in the text. To aid in converting these concentrations, a conversion table is found at the end of this review.

### A. Chemical and Physical Properties

Synonyms: 2-butanone, butan-2-one, ethyl methyl ketone; symbols: MeCOEt, MEK (used in this monograph).

Structural formula:

$$CH_3-CH_2-\overset{\overset{\displaystyle O}{\|}}{C}-CH_3$$

C = 66.63%
H = 11.18%
O = 22.19%

Molecular weight: 72.10.

Boiling range: 78.5°C–90.7°C (boiling point 79.6°C) at 760 mm Hg. Solidifies at -86.35°C.

Specific gravity: 0.8095 at 15.6/15.6°C
0.8051 at 20/15.6°C
0.8068 at 20/20°C.

Viscosity: at 20°C = 0.00441 g/cm-sec, 0.401 centipoise.

Vapor pressure: 40 mm Hg at 6°C
100 mm Hg at 25°C
400 mm Hg at 60°C.

Saturation: approx. 10% (v/v), 295 mg/l, at 25°C.

Soluble: in water (decreasing solubility at higher temperatures); completely soluble in alcohol, acetone, ether, benzene.

Colorless, volatile, inflammable aliphatic ketone with fruity, acetone-like odor.

Distribution coefficient: olive oil/water 0.93 at 30°C.

Thermal stability: Heat of combustion of MEK liquid is 584.17 kcal/g mol wt at 25°C, yielding $CO_2 + H_2O$. Flash point is -4°C (closed cup).

Flammability: Limits for flammability 1.81–9.50% (v/v).

Preparation of MEK: MEK has been produced from calcium acetate and calcium proprionate by rectification of acetone oil during charcoal manufacture. Presumably, the reaction proceeds as:

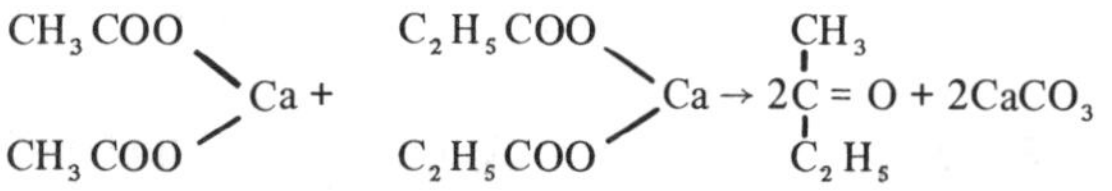

Determination of MEK in gases and biologic fluids: The following methods may be employed.

1. Gas chromatography method – As described by Munies and Wurster,[11] the column is packed with 15% diethylene glycol adipate on 70 to 80 mesh Anakrom ABS (1% boric acid included in the first 6 in.). The column is conditioned at 200°C for 8 to 12 hr, causing retention of MeOH and EtOH on the column, and separation of MEK.

2. 2,4-Dinitrophenylhydrazine method and thin-layer chromatography – This technique has been employed by Buell et al.[12] for incubation mixtures containing one or more carbonyl compounds. Johnson's reagent is added to the mixture and the compounds precipitated as their 2,4-dinitrophenylhydrazones. Fractions soluble in chloroform, *p*-dioxane, and ethanol are obtained after several washings and recrystalizations from ethanol-water. The fractions are separated by TLC and developed according to the following systems: chloroform by benzene: acetate, 4:1; *p*-dioxane by benzene, ethanol, and *n*-hexane, 20:10:1; ethanol fraction the same, but 20:15:1. Identification is by infrared spectroscopy.

3. Iodide titration method – A general method for ketones, the following stoichiometric relationship holds for the reaction:

$$CH_3CH_2COCH_3 + 3I_2 + 3NaOH \rightleftharpoons CHI_3 + CH_3CH_2COOH + 3\,NaI + 2H_2O$$

Recoveries may be influenced (increased) by a secondary reaction:

$$CH_3CH_2COOH + 5I_2 + 4NaOH \rightleftharpoons 2CH_3I + CH_3COOH + 4NaI + 3H_2O$$

As applied to vapors of MEK by Patty et al.,[5] a partially evacuated receiving flask containing 1 *N* NaOH is filled with vapor-air by vacuum displacement, and shaken to absorb the MEK. 0.1 *N* $I_2$ is then added slowly in excess. After 15 min, neutralization and slight acidification is carried out with 2 *N* $H_2SO_4$. The excess iodine is then titrated with 0.05 *N* sodium thiosulfate.

4. Colorimetric Field Test using 2,4-dinitrophenylhydrazine – Smith and Wood[13] have developed a general test for ketone vapor concentration in air that can be employed using ambient temperatures at the sampling site. A measured volume of ketone-air sample is absorbed by a volume of water specific to the particular ketone (20 ml for MEK). One ml of the ketone solution is then transferred to a reaction tube to which 1 ml of 2,4-dinitrophenylhydrazine reagent is added, and the contents divided. Methanolic potassium hydroxide (5 ml) is added, the tube inserted several times, and the solution allowed to stand for approximately 8 min. After transferring to another tube, ketone concentration is determined by color-matching with standard ketone solutions held against a white background.

**B. Medical Uses**

At the present time, MEK has no known medical application. Recent investigations, however, suggest that MEK and related ketones may have some potential therapeutic usefulness as anticonvulsants[14] or local anesthetic agents.[15] Experiments in rats indicate that MEK (80.5 mg/kg) administered orally, prolonged the onset of (but did not protect against) isonicotinic acid hydrazide-induced convulsions, and protected against the maximal electroshock-induced convulsions tested. Acetone (490 mg/kg) appeared to be a better therapeutic agent, however, protecting against the above stimuli as well as pentylenetetrazol-induced seizures, while having a higher $LD_{50}$ than did MEK.[14] In other experiments, a variety of aromatic amino acid compounds with butyl skeletons were tested as local anesthetic agents. Compounds were compared in rabbits against lidocaine or chloral hydrate for local anesthetic actions (topical application to rabbit cornea), as well as for adverse effects ($LD_{50}$ and intradermal injection into rabbit ear). Although MEK derivatives were found as effective in these respects as the known local anesthetics, considerably better results were obtained with butenones and butenols.[15] From the higher toxicity of MEK when compared to similar compounds such as acetone or the butenones, it seems doubtful that MEK would be chosen as a ketone worthy of further exploration for medical application.

### C. Industrial Uses

Reviews of the industrial uses of MEK indicate that in the first half of the present century MEK found wide application as a solvent in processes involving gums, resins, cellulose acetate, and cellulose nitrate, among others.[1,2] These processes included the manufacture of (1) synthetic colorless resins, (2) lacquer, (3) artificial leather, (4) smokeless powder, (5) cosmetics and pharmaceuticals, and (6) lubricating oils. MEK was also used extensively in the synthetic rubber industry, particularly in the manufacture of waterproofing material.[3] New applications have since been found for this substance, (1) in the production of paraffin wax[16] and high-grade lubricating oil[17] from crude oil, (2) as a solvent in various stages of plastic manufacture, including polymerization (e.g., of styrene,[18] vinyl butyrate,[19] and vinyl chloride[20]), as a blowing agent in the extrusion of thermoplastic polymers,[21] in removing molded plastic from casts,[22] and in cleaning polymerization reactors.[23]

### D. Consumer Products

MEK is found in several household products, such as surface coating (lacquer and varnishes),[1,2] paint remover,[2] and glues.[4] Gleason[6] reports the following commercial products containing MEK.

1. Aerosol products: component of volatile vehicle; coating for aluminum and chrome (one product, volatile vehicle 48%, of that 32% MEK); hammer finish and spray paint (one product, volatile vehicle 34%, of that, 10% MEK); spray paint (one product, volatile vehicle 46%, concentration of MEK unknown); epoxy spray paint (three products, volatile vehicle 57%, 51%, 49%; of the volatile vehicles, 41%, 42% and 60% MEK).
2. Adhesive and cements: fifteen products (< 25%, 15%, 90% [including toluol]; remaining, concentration unknown).
3. Sealers: waterproofing (one product, concentration unknown); wood sealer (one product, 15%); window sealer (one product, concentration unknown); liquid plastic (one product, concentration unknown).
4. Liquid solder: three products (< 90%, < 15% and unknown concentration).
5. Other products: primer (one product, concentration unknown); brush cleaner (one product, concentration unknown); mortar powder (one product, concentration unknown).
6. Catalyst plastic filler hardener: twelve products (ten with 60%; two, concentration unknown).

### E. Human Investigation

An absence of documented fatalities due to MEK exposure has led to the belief that this solvent is relatively safe. Nonetheless, occupational hazards have been reported since the early days of its use, and, in turn, have stimulated further investigation in man. Despite substantial gaps, such as in studies of MEK blood levels during vapor inhalation or human metabolism of MEK, the literature contains some important pharmacodynamic and toxicologic studies of MEK in man.

#### *1. Occupational Hazards*

The most common complaint from occupational exposure to MEK is dermatitis from direct skin contact with the liquid, as well as from exposure to the vapors.[3] Nelson and co-workers[7] found that the opportunity for more serious physiologic consequences may be lessened by the inability to tolerate the odor and mucous membrane irritation from MEK vapors greater than 200 ppm (0.59 mg/l) for an 8-hr working day. Prior to the settling of a TLV at this level,[24] exposures to higher vapor concentrations had been implicated as the cause of more severe symptoms. In 1944, Smith and Mayers[3] reported that workers handling MEK while manufacturing raincoat waterproofing material in a poorly ventilated factory developed severe dermatosis with a complete lack of sensation in the digits and limbs. Analysis of the environmental conditions revealed the presence of 300 to 600 ppm (0.9 to 1.8 mg/l) MEK vapor. Additionally, central nervous system depression and narcosis, including acute loss of consciousness, occurred in two women in a similar plant in which ketone vapors (MEK and acetone) were assessed at 1000 ppm.[3] Workers using products containing MEK in confined areas may also be subject to vapor-induced toxicity. Such a possibility was implicated in a more recent account of retrobulbar neuritis in a man removing paint from an airplane hangar. Although the author admits lack of proof that MEK was the cause, dizziness, nausea, and the finding of steadily diminishing concentrations of methanol (a possible metabolite of MEK) in the blood during hospitalization suggested that MEK might have induced the ocular damage.[25]

*2. Addiction and Abuse*

Abuses of industrial solvents, such as "glue sniffing" or other purposeful inhalation of solvent vapors, became popular in the early 1960s, particularly with young children and teenagers.[4,8,9] MEK is one of the substances, in combination with other solvents such as toluene, benzene, and acetone, encountered by those intentionally seeking the narcotic effects of glues.[4] In a review of the physiologic and toxicologic effects following inhalation of volatile substances (particularly toluene), Press and Done[4] considered the narcotic effects of acetone, and presumably MEK, to be considerably greater than ethanol. They concluded, however, that its popularity is reduced by irritating effects on the eyes and nose.

*3. Other Sources of Exposure*

MEK has been detected, along with various other organic substances, in the gas phase of cigarette smoke collected from a standardized smoking machine. Infrared spectrophotometric analysis of the gases from several tobacco blends showed that the fast-burning "burley" and 50:50 "burley/bright" mixture yielded 1.8 to $2.3 \times 10^{-2}$ ml/puff (representing 6 to $7 \times 10^{-2}$ mol % of the total volume) while the slower-burning "bright" tobacco produced only about 25% of this amount. Similar burley versus bright differences were obtained for other carbonyl compounds. About three fold as much acetone was formed as MEK in this investigation.[26]

MEK has also been detected in the exhaled air[27] and in the urine of apparently healthy individuals[28] as well as diabetics.[29] Whether these findings reflect inadvertent contact with a precursor substance or MEK itself, or whether MEK may be a metabolite of a normal biochemical process, perhaps, as suggested, isoleucine catabolism,[28] is unknown. However, gas chromatography and mass spectroscopy have recently shown that MEK is a natural component of the flavor and/or odor of a wide range of foods, including milk and cream,[30] ripening blue-veined cheese,[31] tomatoes,[32] cottonseed oil,[33] roast barley,[34] degraded starch,[35] honey,[36] fresh white bread,[37] fish (cod),[38] cooked turkey,[39] potato chips,[40] mushrooms,[41] coffee,[42] chicken,[43] eggs,[44] and whiskey.[45] Of additional interest, carbonyl compounds, including MEK, have been found in the protein fraction of enzyme digests from rabbit lungs extracted following in vivo exposure to as little as 1 to 5 ppm of ozone.[12]

*4. Hazards in the Operating Room*

In the early part of the century, MEK was found to be one of the impurities in anesthetic ether. When tested by inhalation in dogs lightly anesthetized with ether, up to 5% (147.5 mg/l) MEK, as well as a similar concentration of acetone or di-ethyl ketone, did not appear to affect respiration or blood pressure greater than did ether alone.[46,47] In considering these findings, it should be pointed out that the concentrations were estimates based on application of the liquids by a rag over the mouth and nose and, therefore, may inaccurately reflect the true concentration inspired.

*5. Experimental Inhalation*

In 1935, Patty and co-workers[5] momentarily exposed men to graded concentrations of MEK vapor in air. They reported that 0.33% (9.7 mg/l) MEK vapor produced a moderately offensive odor and irritation of the nose and eyes, while 3.3% (97.4 mg/l) to 10% (295 mg/l) was intolerable even for a brief exposure. It was pointed out that the odor and irritation noticed with 0.33% (9.7 mg/l) occurred at about one sixth the minimum flammable concentration of 2% (59 mg/l), thus serving as a good warning sign for fire hazards. In a later study, Nelson et al.[7] reported that of ten subjects tested, some found 100 ppm (0.295 mg/l) to produce slight nose and throat irritation in 3 to 5 min, while 350 ppm (1.03 mg/l) irritated the eyes, nose, and throat of the majority. Most subjects estimated that 200 ppm (0.59 mg/l) would be the maximum concentration that would be tolerable over an 8-hr work day. (A similar value was estimated for acetone, although on acute exposure, a concentration of 500 ppm irritated the mucous membranes of most subjects). The 8-hr estimate for MEK concentration in the workroom (200 ppm or 0.59 mg/l) has since been part of the basis for the current TLV for MEK of 200 ppm.[24] Even at this level, however, Nakaaki[48] recently found effects on mental acuity, as tested by the ability to estimate the passage of time. Two men and two women were exposed to MEK, 90 to 270 ppm (0.27 to 0.81 mg/l), for 30, 60, 90, and 120 min in the morning and afternoon (total 5 hr). The author reports that the men underestimated the elapsed time in a time estimation test, while the mean value for women was about the same as controls, but there was a two- to threefold increase in deviation. By contrast, for acetone (170 to 690 ppm), both men and women overestimated the time test.

*6. Absorption*

Absorption by the skin represents another major route for MEK to enter the body. Studies using a percutaneous absorption cell tightly fitted to the skin, indicate that the degree of hydration of the skin is one of the most important factors in influencing absorption by this route.[11,49] As measured by the concentration of MEK vapors collected in exhaled air at various times following exposure, with wetted skin there was an increase in the normal absorption, whereas absorption was delayed in dessicated skin. These differences can be important, since Malten et al.[50] have evidence that MEK causes damage to the stratum corneum (horny layer), with consequent loss of water from deeper layers. They suggest certain preventative precautions, such as pretreatment of the skin with petrolatum as well as splash-preventative measures.

*7. Urinary Excretion*

To our knowledge, there is no available data concerning blood levels or metabolism of MEK in man. However, ketone has been found in the urine after inhalation of MEK or acetone. As measured by the 2,4-dinitrophenylhydrazine colorimetric technique, 8 to 12 mg of ketone were detected within an 8-hr period following inhalation by two subjects of 110 to 248 ppm (0.32 to 0.74 mg/l) of MEK. MEK has also been detected in the urine of presumably healthy, unexposed people. In the latter finding, it was hypothesized that MEK could be a decarboxylation product of α-methylaceto-acetate derived possibly from the catabolism of isoleucine.[28] Current knowledge, however, suggests that the ketone may have been derived from food.[30-45]

*8. Excretion by the Lungs*

MEK has been detected in the expired air following absorption through the skin, oral ingestion in encapsulated form,[11,49] and inhalation of vapor.[51] Munies and Wurster[11] found exponentially decreasing concentrations of MEK in air when monitoring began 30 min after ingestion of 375 mg of MEK in capsule form. The peak concentration in the expired air was 200 mg/l, which fell to 4 mg/l after 270 min. When MEK was applied to the skin by a cell attached to the forearm, ketone in the expired air increased in concentration, reaching a steady state level of 5 mg/l in 60 min. Pulmonary excretion of ketone following MEK vapor inhalation has been measured by Tada and associates[51] using gas chromatographic techniques. After 2 hr exposure, there was roughly a linear relationship between the inspired concentration of 50 to 350 ppm (0.15 to 1.03 mg/l) and the expired concentration, which decreased 15 min and 30 min after the inhalation period. Representative expired values after inhalation of 200 ppm (0.59 mg/l) MEK were 15 ppm (0.045 mg/l) immediately after exposure, 10 ppm (0.03 mg/l) after 30 min, and 8 ppm (0.024 mg/l) after 60 min.

**F. Metabolism and Disposition in Animals**

In the early part of this century, Cushny[52] observed by iodometric techniques that a 3 kg cat injected intravenously with 0.16 g MEK exhaled only about 5% of the dose (9.1 mg) within the first 90 min. Comparing this pulmonary elimination rate to the rates of several other volatile solvents (e.g., chloroform 22% in 60 to 90 min), he concluded that elimination via the lungs is passive, generally governed by the oil:water partition coefficient of the substance. Another report indicated that MEK is also catabolized by tissues, including slices of liver, spleen, kidney, lung, and muscle. MEK catabolism was found greater than for acetone.[53] Recently, Liebman[54] reported that ketones are reduced to corresponding alcohols, e.g., butan-2-one (MEK) to butan-2-ol. These reactions are mostly confined to the cytosol rather than microsomal cell fractions of liver and other organs. Aliphatic ketones such as acetone and MEK oxidize about equally the cofactors NADPH and NADH, while ketones with aromatic groups, for example acetophenone and metyrapone, do not oxidize NADH more than about 10% that of NADPH. The rat liver ketone reductase system was not able to be induced with phenobarbital.

**G. Toxicologic Investigation**

In this section, experiments designed to determine mortality and other toxic effects of MEK are presented with respect to route of administration, acute or chronic administration, and species studied.

*1. Oral Administration*

Kimura and co-workers[55] carried out acute oral toxicity experiments in rats of various age groups: 14-day-old, young adult, and older adult. The $LD_{50}$ (95% confidence limits) for a single dose of MEK was found to be 3.1 to 3.6 ml/kg (2511 to 2916 mg/kg), varying very little among age groups. Acetone appeared somewhat less toxic in young and old adults ($LD_{50}$: 9.1 ml/kg and 8.5 ml/kg, respectively), but very similar to MEK in 14-day-olds ($LD_{50}$: 5.6 ml/kg). In addition, the author suggested "maximal permissible limits" for acetone (0.005 ml/kg) and MEK (0.005 ml/kg), based on 1/1000th the dose at which the first observable gross symptom occurred in young adults.[55]

*2. Parenteral Administration*

Divencenzo and Krasavage[56] administered MEK to guinea pigs by intraperitoneal injection, in order to determine liver toxicity, as reflected in elevated levels of serum ornithine carbamyl transferase activity. They found that 750 mg/kg caused no discernable increase from controls, whereas there was a tripling of enzyme activity following 1500 mg/kg. After 2000 mg/kg, one (of four) guinea pigs died, while serum activity in the remaining animals was about six times that of controls. Changes were also seen in histological studies, with the appearance of "moderate" lipid deposition in hepatocytes. However, tissue sections stained with hematoxylin and eosin showed no apparent hepatocellular damage. By comparison with the "highly toxic" halogenated hydrocarbon solvents such as carbon tetrachloride and chloroform, MEK was found to have "low liver toxicity," based on an intraperitoneal dose greater than 500 mg/kg required to elevate serum ornithine carbamyl transferase activity.

*3. Inhalation Toxicity: Single Exposure in Mice*

La Belle and Brieger[57] determined the saturated vapor toxicity of MEK and other solvents in mice, using the geometric mean survival time during exposure as the index of relative toxicities of these compounds. They report that mice exposed to saturated MEK vapor (nominal concentration 103,000 ppm, or 304 mg/l; temperature not reported) had a mean survival time of 43 min. When MEK was the major constituent of a test solvent mixture containing propylene formal, 1,3-amylene oxide, tetrahydrofuran, 1,3-butylene oxide, dimethyl acetal, propylene acetal, and acetone, the mean survival time was 59 min. This was closer to the time for MEK than for any of the other mixed solvent constituents.

*4. Inhalation Toxicity: Single and Chronic Exposure in Rats*

La Belle and Brieger[57] report that rats exposed to MEK vapor for 4 hr exhibited narcosis and died quickly thereafter. They found the $LC_{50}$ (concentration in ppm x mean survival time in min/60) of 11,700 ± 2400 standard deviation for MEK alone and 10,000 ± 2100 for the solvent mixture, described above, in which MEK was the prime constituent. MEK was also tested in rats by Carpenter et al.[58] for the purpose of ranking its acute toxicity with those of 95 other solvents. Concentrations of each compound were increased by a factor of two until reaching the level that killed 2 of 6, 3 of 6, or 4 of 6 rats in 4 hr. MEK was classified as "moderately hazardous," killing two to four rats at a concentration of 2000 ppm (5.9 mg/l). Acetone was not tested.

In chronic experiments, La Belle and Brieger[57] administered MEK alone (235 ppm ± 26, 0.693 to 0.077 mg/l) and in the mixture of solvents tested in the last section (226 ppm ± 25 total concentration) for 12 weeks (8 hr/day, 5 days/week). All rats survived. Hematologic as well as gross and microscopic pathologic examination revealed no significant differences from controls. The solvents mixture was found not to have a potentiating or synergistic effect on MEK toxicity. A somewhat different type of study was carried out to investigate in rats the development of peripheral neuropathy among users of printer's ink. Methyl *n*-butyl ketone was tested for six weeks (8 hr/day, 5 days/week) both alone (200 ppm) and in combination with MEK (2000 ppm or 5.9 mg/l). All rats surviving the duration of the experiment exhibited muscular weakness of the limbs, which with methyl *n*-butyl ketone alone disappeared 3 hr after daily exposure, but with the combination, remained at least 24 hr. After sacrifice, microscopic examination of the sciatic nerve revealed extensive axonal damage as well as widespread perinodal and segmental deterioration of myelin. Although the presence of MEK did not seem to be a requirement for the neuropathologic effects, the role of MEK was to be further investigated.[59]

*5. Inhalation Toxicity: Single and Repeated Exposure in Guinea Pigs*

In 1935, Patty and co-workers[5] carried out a thorough investigation of acute toxicity in guinea pigs, comparing the duration and concentration of MEK exposure and graded signs of toxicity: irritation of nose and eyes (rubbing nose, squinting), lacrimation, incoordination, narcosis, gasping-type respiration, and finally, death. Highlights of their results are as follows: no symptoms in 810 min with 0.33% (9.7 mg/l) MEK; nose and eye irritation, lacrimation in 2 to 4 min, incoordination in 40 min, narcosis in 90 min with 1% (29.5 mg/l) MEK; gasping in 180 min, death in 200 to 260 min with 3.3% (97.4 mg/l) MEK; death in less than 60 min with 10% (295 mg/l) MEK. Postmortem examination revealed hemorrhage, congestion, and edema of vital organs.

Guinea pigs were also used in the chronic inhalation experiments by La Belle and Brieger[57] discussed in Sections G.3 and G.4. As in the case with the rats, these animals were asymptomatic with MEK (235 ± 26 ppm or 0.693 ± .077 mg/l), as well as mixed solvent (226 ± 2.5 ppm). In general, they did not prove as hearty as the rats, however, since two control and two experimental animals fell sick and died, presumably from vitamin deficiency and poor nutrition.

*6. Inhalation Toxicity: Single and Repeated Exposure in Dogs*

In Section E.4, it was noted that MEK vapor, reported to be 5% (147.5 mg/l), was tested in lightly etherized dogs for respiratory and blood pressure effects. Although the stated concentration of MEK (as well as of other compounds studied) is questionable due to the method of administration, it is interesting that no further respiratory or blood pressure effects were observed over those produced by ether alone.[46,47]

In a chronic inhalation study, dogs exposed to MEK vapor exhibited muscular disability, vomiting, and emaciation. In addition, ammonium magnesium phosphate crystals were found in the urine, but disappeared when milk was substituted for meat in the diet.[60]

**H. Pharmacodynamics**

The purpose of this section is to define the toxicity of MEK by its effects on individual organs and organ systems.

*1. Central Nervous System Toxicity*

Central nervous system involvement in MEK intoxication has been shown by experimentation in guinea pigs,[5] rats,[57] and man,[48] as well as by documented reports of industrial accidents[3] and intentional abuse of glues.[4,8,9] In guinea pigs, Patty and co-workers[5] found that effects are related to the concentration and duration of exposure. Two stages of toxicity, incoordination and narcosis, were clearly discernable. Incoordination appeared after breathing 1% (29.5 mg/l) MEK for 90 min, 3.3% (97.4 mg/l) for 18 to 30 min, or 10% (295 mg/l) for 3 to 5 min. Narcosis, defined as loss of consciousness, occurred later and/or at higher concentrations: 90 min with 1% (29.5 mg/l), 48 to 90 min with 3.3% (97.4 mg/l), or 10 to 11 min with 10% (295 mg/l) MEK. Neither incoordination nor narcosis could be observed with 0.33% (9.7 mg/l), even after the maximum duration tested, 810 min. Narcosis developed quickly in rats exposed to 7,850 to 20,200 ppm (23.2 to 59 mg/l), with an $LD_{50}$ reported of 11,700 ppm (34.5 mg/l).[57]

Human experimentation suggests that values for concentrations affecting the central nervous system of animals may be too high to serve as useful estimates in man. This is shown, for example, in the mental acuity test carried out by Nakaaki,[48] in which deficiencies in perception of time duration were noted in 30-min test intervals during exposure to only 90 to 270 ppm (0.27 to 0.81 mg/l) MEK over a 4-hr period. This evidence is further supported by accounts of acute poisoning (including unconsciousness) in two young women workers inadvertently exposed to 398 to 561 ppm (1.2 to 1.7 mg/l) MEK for only a short period of time.[3]

The deliberate sniffing of glues, solvents, and solvent mixtures containing MEK and other compounds is a common abuse of these materials by children and teenagers, especially those seeking a narcotic effect.[4,8,9] In addition to dizziness, adverse effects of solvent sniffing have been reported to include stupor, sleepiness, loss of consciousness, excitability, nausea, and vomiting.[8]

*2. Peripheral Nervous System Toxicity*

In their report on problems of industrial exposure to MEK, Smith and Mayers[3] state,

> . . . workers developed disabling dermatosis and several workers complained of numbness of the fingers and arms. One worker described one shoulder as feeling 'as though it was made of dough and did not belong to him' so that he could not lie on it at night. Another worker complained of numbness in the legs and a tendency for them to suddenly give way under him.

These toxic effects occurred in an environment containing 300 to 600 ppm (0.89 to 1.8 mg/l) MEK. Additional symptoms of ketone exposure relating to the peripheral nervous system have been reported from experiments in dogs,[60] rats,[59] and rabbits.[61] Dogs chronically exposed to MEK were described as developing muscular disability.[58] Duckett and co-workers[59] investigated ketone toxicity, mainly methyl *n*-butyl ketone, in rats. While neurophysiologic toxicities appeared to be greater with chronic administration of methyl *n*-butyl ketone and MEK in combination rather than with methyl *n*-butyl ketone alone, (e.g., muscular weakness of limbs for 24 versus 3 hr after daily administration), histologic changes were roughly the same as for methyl *n*-butyl ketone alone. The finding of hypertrophy of axons of the sciatic nerves and secondary breakdown of myelin suggested early changes of neuropathy.[59] Actions on peripheral nerve transmission have been shown by Gautier et al.[15] in a rabbit cornea preparation treated with aromatic amino alcohols and ketones containing a butyl chain. Substituted MEK was found to be fairly active as a local anesthetic agent in this preparation.

*3. Dermatotoxicity*

Dermatitis is the most common toxicologic effect of MEK in industry, occurring from contact with either the liquid or the solvent vapor.[3] Experimentally, skin contact with the liquid (15 min/day for 6 days) caused destruction of the horny layer (stratum corneum) in man, as shown by Malten and associates.[50] Water loss was minimized, however, by formation of a temporary protective barrier. Although toxic, MEK was found less hazardous than rectified light benzene or dimethyl sulfoxide.[50] Application of MEK to rat abdominal skin caused hemoglobinuria within 20 min.[61]

*4. Oculotoxicity and Nasal Toxicity*

Men exposed to 0.33% (9.7 mg/l) MEK found that the vapors were moderately irritating to the eye and had a moderate to strong odor, while the eye and nose irritation as well as the odor from 1% (29.5 mg/l) were not tolerable after a few breaths. By contrast, guinea pigs apparently withstood 0.33% (9.7 mg/l) for up to 14 hr (the duration of the experiment) without any signs of nose or eye irritation.[5] A single case of retrobulbar neuritis was suspected of being caused by MEK in paint remover, but this could not be proven.[25]

Investigations using various rabbit eye preparations indicate that MEK and other industrial solvents can cause miosis[62] and conjunctival edema.[63] Opacity of the cornea occurred in guinea pigs exposed to 10% (295 mg/l) vapor for 30 min. However, in those guinea pigs that recovered, the eyes also appeared normal after 4 to 8 days.[5]

*5. Pneumotoxicity*

Patty and co-workers[5] noted gasping-type respiration in guinea pigs exposed to 3.3% (97.4 mg/l) MEK vapor for 180 min or 10% (295 mg/l) vapor for 20 to 30 min. This effect of MEK occurred in animals that were already unconscious, and was the last discernable toxicity prior to death of the animal. Lungs of pigs which died from the inhalation of MEK were markedly congested and emphysematous. The cause of death in these experiments, in fact, might have been construed as pneumotoxicity, except that all animals regaining consciousness survived (i.e., narcosis rather than pneumotoxicity seemed to be more related to death during MEK inhalation). By contrast to these results in guinea pigs, a study in dogs revealed no effect of 5% (147.5 mg/l) MEK on respiration during light ether anesthesia.[46,47] However, the given concentration is open to question.

*6. Cardiovascular Toxicity*

As noted previously, to the best of our knowledge, the only account of MEK effects on the cardiovascular system is an early study in lightly anesthetized dogs. A nominal concentration of 5% (147.5 mg/l) had no effect on blood pressure greater than with ether alone.[46,47]

*7. Hepatotoxicity*

Schwetz and co-workers[64] report that 1,000 to 3,000 ppm (2.95 to 8.9 mg/l) MEK vapor did not affect absolute or relative liver weight, or the gross appearance of the liver in chronic inhalation studies with pregnant rats (7 hr/day from day 6 to

15 of gestation). Likewise, serum glutamic oxaloacetic transaminase activity was not affected. By comparison with these findings, carbon tetrachloride (300 to 1,000 ppm) caused mottling of the liver, an increase in liver weight, and an increase in serum enzyme activity. It was concluded that the nonchlorinated MEK had little or no hepatotoxicity as compared with carbon tetrachloride. The low hepatotoxicity of MEK as compared with halogenated hydrocarbons was later confirmed by Divencenzo and associates[56] in guinea pigs given a single i.p. injection. Measuring a serum enzyme activity they believed more specific than glutamic oxaloacetic transaminase for the liver, i.e., ornithine carbamyl transferase, these investigators detected liver damage following administration of 1,500 mg/kg but not 750 mg/kg of MEK. Fatty deposition, but no hepatocellular damage, could be seen after 1,500 mg/kg or 2,000 mg/kg. Even greater damage, however, could be seen following administration of only 25 mg/kg carbon tetrachloride.

*8. Other Organ Toxicities*

Gross pathologic examination of guinea pigs that died during exposure to 3.3% (97.4 mg/l) or 10% (295 mg/l) MEK revealed slight congestion of the brain and marked congestion of the systemic organs.[5] Formation of large quantities of urinary magnesium ammonium phosphate crystals in dogs chronically exposed to MEK vapor has been noted by Verstraete and co-workers.[60] Free ammonia appears to be necessary for these crystals, since replacing meat with milk in the diet abolished formation of the crystals while preventing the increase in $NH_3$ in meat-fed dogs.

*9. Embryotoxicity and Fetotoxicity*

Embryotoxicity and fetotoxicity of MEK were studied in pregnant rats subjected to inhalation of 1,000 and 3,000 ppm (2.95 and 8.9 mg/l) MEK vapor for 7 hr/day from days 6 to 15 of gestation. Schwetz et al.[64] report finding some retarded fetal development, such as in body weight and/or size, with MEK (300 and 1,000 ppm or 0.89 to 7.95 mg/l), carbon tetrachloride (300 and 1,000 ppm), but little with 1,1-dichloroethane (3,600 and 6,000 ppm). MEK was most embryotoxic, causing several anomalous skeletons, acaudia, imperforate ani, and brachygnathous fetuses. However, relative to their TLV, all of these substances were relatively non-embryotoxic, especially as compared to chloroform.

*10. Cytotoxicity*

MEK was found "moderately toxic" to Ehrlich-Landschutz diploid ascites tumor cells. Cells taken from MNRI mice were prepared for incubation, and then incubated for 5 hr with various solvents added to sealed tubes containing the incubation mixtures. While most solvents did not cause injury to the cells, some, such as MEK, benzyl chloride, carbon tetrachloride, and 1,1,1,2-tetrachloroethane (all in concentrations of 100 ppm) caused injury to 10 to 20% of the cells. Others were even more toxic, e.g., a large number of injured cells after 3.5 hr with 5 ppm acrolein.[65] MEK diperoxide has also been reported to sensitize rats with lymphosarcoma to the effects of X-irradiation.[66] When compared with solvents having longer side chains, MEK had low cytotoxicity against *S. typhosa* and *Saccharomyces cerevisiae.*[67]

**I. Clinical Toxicology**

No deaths have been reported thus far from MEK intoxication. Gleason et al.[6] consider MEK to be "moderately" toxic, based on a probable human lethal dose of 5 to 15 g/kg. They note, however, as had been previously hypothesized by Nelson and co-workers,[7] that the irritating odor normally prevents exposure to concentrations which might be expected to have significant physiologic effects. Smith and Mayers[3] note that most complaints about MEK exposure concern dermatitis and debilitating dermatosis among workers using the solvent. However, they also recorded two cases of acute poisoning in two young women (18 and 19 years old) working in a poorly ventilated room where MEK concentration in the air was measured as 398 to 561 ppm (1.2 to 1.7 mg/l). Acetone was also present at concentrations of 330 to 495 ppm. The first woman noted gastric distress shortly after beginning work one morning, and was observed by a co-worker to have watering eyes. She became unconscious in the rest room that afternoon. At the hospital, she was admitted in a coma, with acetone on the breath, but negative results of a urine test precluded diabetes. Occasional twitching of the face and extremities, as well as hyperactive reflexes, were present at physical examination. Other physical findings were pulse 112/min, temperature 98°F, blood pressure 128/70 mm Hg, and high spinal fluid pressure. Treatment with nikethamide (6 ml, i.v.) resulted in immediate recovery, with only a severe headache remaining. The next day the headache subsided, and she was discharged 1 day later. At about the same time as the first incident, a second woman worker fainted at work, followed by convulsions and regaining of consciousness. She was taken to the same hospital, appearing mentally confused and complaining of headache. She was not treated, but allowed to go home. She resumed work two days later. A third case of possible MEK poisoning has more recently been reported by Berg,[25] concerning the development of retrobulbar neuritis in a man removing paint from an airplane hanger using solvent containing MEK. Although early signs included headache, mild vertigo, and blurred vision, within a short period (90 min) only blurred vision persisted. Ophthalmologic examination indicated retrobulbar neuritis. The urine was positive for methanol and formaldehyde (possible metabolites of MEK), although the patient denied ingesting any toxic fluid or drugs. Treatment was begun with 2 ml vitamin B complex, 1 mg vitamin $B_{12}$ and 80 mg oral prednisone. After 36 hr, vision became normal. The author speculates that methanol, when mixed with other metabolites of MEK, may assume a more toxic effect on the visual apparatus than methanol alone.

### J. Threshold Limit Value

The TLV for MEK is 200 ppm (0.59 mg/l) at 25°C, 760 mm Hg.[24] This represents approximately 590 mg of particulate matter per cubic meter of air. The TLV value is based on the significant clinical findings of Smith and Mayers[3] at 300 to 600 ppm (0.89 to 1.8 mg/l), as well as the report of Nelson et al.[7] on the estimated maximal concentration (200 ppm, 0.59 mg/l) considered tolerable to the eyes, nose, and throat of subjects for an 8-hr period.

### K. Conversion Table for Concentrations of MEK

The following table is presented in order to aid in converting MEK concentration in volume/volume to weight/volume. Since the molecular weight of MEK is 72.10, and the volume occupied by one molecule weight at 25°C is 24.45 liter, 2.95 g occupy 1 liter.

| mg/l | ppm | v/v% |
|---|---|---|
| 0.00295 | 1 | |
| 0.0295 | 10 | 0.001 |
| 0.097 | 33.0 | |
| 0.295 | 100 | 0.01 |
| 0.97 | 333 | |
| 2.95 | 1,000 | 0.1 |
| 9.7 | 3,334 | |
| 29.5 | 10,000 | 1.0 |
| 97.4 | 33,344 | |
| 147.5 | 50,000 | 5.0 |
| 295 | 100,000 | 10.0 |

Chapter 7

# ACUTE ORAL, INTRAPERITONEAL, AND INHALATIONAL TOXICITY IN THE MOUSE

As described in Chapter 6, the acute toxicity of MEK has been determined in a variety of species (mouse,[57] rat,[55,57,58] guinea pig[5,56]) using several routes of administration (oral,[55] intraperitoneal,[56] and inhalational[5,57,58]). No reports are known to us, however, in which the acute toxicities by these routes have been compared in a single species. The present investigation, therefore, was undertaken using the mouse, to reexamine the acute toxicity of MEK following oral, intraperitoneal, and inhalational administration.

## A. Methods

Mice (male, CF-1 strain, Charles River Laboratories, Wilmington, Massachusetts) were divided into three groups for oral, intraperitoneal, and inhalational studies. MEK, certified laboratory and manufacturing grade, was obtained from Fisher Scientific Co., Fair Lawn, New Jersey.

### *1. Oral Toxicity*

Mice, weighing 25 to 31 g, were given MEK by the oral route using a gastric tube. Following pilot experiments to establish a reasonable dose range, six groups of ten mice were given doses of 2.0, 2.5, 3.0, 3.7, 4.2, and 5.1 g/kg.

### *2. Intraperitoneal Toxicity*

Five groups of ten mice each (weight range: 19 to 22 g) received MEK by intraperitoneal injection. On the basis of preliminary experiments to establish an effective dose range, doses of 0.5, 0.75, 1.0, 1.5, and 2.0 g/kg were administered.

### *3. Inhalational Toxicity*

Six groups of ten mice (weights: 20 to 23 g) received MEK by inhalation for 45 min. Vapor concentrations of MEK were produced by volatilizing MEK liquid into a stream of charcoal-filtered air, directed at a known rate into a gas bag. The contents of the bag were then allowed to empty at a rate of 2 to 3 l/min into a sealed glass chamber containing the mice. Constant vapor flow in the chamber was achieved by applying a small vacuum at the exit port in the chamber. An initial trial was carried out to determine the exposure time causing 100% mortality at the maximum (saturated) MEK concentration of 10% (295 mg/l). This duration, 45 min, was then used as the end point at which percent mortality was determined for subsequent concentrations (v/v%): 5% (147.5 mg/l), 6% (177.0 mg/l), 7% (206.5 mg/l), 8% (236.0 mg/l), and 9% (265.5 mg/l).

### *4. Calculation of $LD_{50}$*

The criterion for acute toxicity by the oral and intraperitoneal routes was survival up to 24 hr following administration. For inhalational studies, survival was observed directly following 45 min of exposure. Data (percent mortality) were plotted and $LD_{50}$ (95% fiducial limits) were determined according to the probit method.[68]

## B. Results

The progression of toxic effects following administration of MEK was observed to be incoordination, narcosis (unconsciousness), respiratory depression, and death. Results of acute toxicity experiments in which MEK was administered orally are shown in Table 7.1. There were no deaths 24 hr following 2.0 g/kg, while all mice receiving 5.1 g/kg died within this period. The oral $LD_{50}$ (95% fiducial limit) was found to be 3.14 ± 0.67 g/kg.

In Table 7.2 percent mortality 24 hr following intraperitoneal administration is presented. It can be seen that 0.5 g/kg caused no mortality, while eight of ten mice were killed with 2.0 g/kg. The $LD_{50}$ for the intraperitoneal route was 1.66 ± 0.74 g/kg.

In initial inhalation experiments, the saturated vapor (10%, 295 mg/l) caused 100% mortality within 45 min; thereafter, this period of exposure was used for all succeeding concentrations. Table 7.3 indicates that 9% also killed all mice, whereas 5% caused no mortality. The $LC_{50}$ by the inhalational route for 45 min was determined to be 6.95 ± 1.10% (205.0 ± 32.5 mg/l).

## C. Discussion

The acute toxicities of MEK in mice were determined in this series of experiments. Administration by the oral route revealed an $LD_{50}$ of 3.14 ± 0.67 g/kg (95% fiducial limit, Table 2.1). Although species differences make it difficult to compare the present data with previous studies,

Kimura and associates[55] obtained similar results in the rat, reporting an oral $LD_{50}$ of 3.1 to 3.6 ml/kg (2.5 to 2.9 g/kg).

The $LD_{50}$ in mice given MEK by the intraperitoneal route was found to be 1.66 ± 0.74 g/kg (Table 7.2). No other $LD_{50}$ values for this route have been reported in the literature, although Divencenzo and associates[56] noted the death of one of four guinea pigs following intraperitoneal administration of 2.0 g/kg.

Inhalation experiments indicated that exposure to 10% (295 mg/l) and 9% (265.5 mg/l) MEK caused 100% mortality after 45 min of exposure (Table 7.3). This value is close to the 43-min mean survival time in mice exposed by La Belle and Brieger[57] to saturated vapor, since almost all mice in the current investigation died near the end of the exposure time. As for results in other species, Patty and co-workers[5] reported a similar value for average survival time (45 to 55 min) in guinea pigs exposed to 10% MEK. The pattern of onset of toxicity, although more easily observed in guinea pigs than in mice, had roughly the same order, including incoordination, narcosis (unconsciousness), respiratory depression, and death.

An $LC_{50}$ of 6.95 ± 1.10% (205.0 ± 32.5 mg/l) was found for MEK administered by inhalation in mice (Table 7.3). In longer exposures than those currently reported, Carpenter et al.[58] found 33 to 66% mortality in rats after 4 hr exposure to 2000 ppm (0.2% or 5.9 mg/l). In contrast, La Belle and Brieger[57] required 11,700 ppm (1.2% or 34.5 mg/l) for 50% mortality in 4 hr. Reconciliation of these two reports, as well as meaningful comparison with the present results in the mouse, is difficult.

### D. Summary

The acute toxicity of MEK administered by the oral, intraperitoneal, and inhalational routes has been determined in the mouse. Results indicate an oral $LD_{50}$ (95% fiducial limit) of 3.14 ± 0.67 g/kg and an intraperitoneal $LD_{50}$ of 1.66 ± 0.74 g/kg. For the inhalational route, an $LC_{50}$ of 205.0 ± 32.5 mg/l (6.95 ± 1.10%) for 45 min was found.

TABLE 7.1

**Acute Oral Toxicity of MEK in Mice**

| Group* | Dose (g/kg) | % Mortality | $LD_{50}$ (g/kg) 24 hr | Regression coefficient |
|---|---|---|---|---|
| 1 | 2.0 | 0 | | |
| 2 | 2.5 | 20 | | |
| 3 | 3.0 | 40 | 3.14 ± 0.67 | 0.972 |
| 4 | 3.7 | 70 | | |
| 5 | 4.2 | 90 | | |
| 6 | 5.1 | 100 | | |

*Each group, in acute toxicity experiments, contained ten mice.

TABLE 7.2

**Acute Intraperitoneal Toxicity of MEK in Mice**

| Group | Dose (g/kg) | % Mortality | $LD_{50}$ (g/kg) 24 hr | Regression coefficient |
|---|---|---|---|---|
| 1 | 0.5 | 0 | | |
| 2 | 0.75 | 10 | | |
| 3 | 1.0 | 20 | 1.66 ± 0.74 | 0.942 |
| 4 | 1.5 | 30 | | |
| 5 | 2.0 | 80 | | |

TABLE 7.3

**Acute Inhalational Toxicity of MEK in Mice**

| Group | Concentration | | % Mortality | $LC_{50}$ (mg/l) 45 min | Regression coefficient |
|---|---|---|---|---|---|
| | % v/v | mg/l | | | |
| 1 | 5 | 147.5 | 0 | | |
| 2 | 6 | 177.0 | 20 | | |
| 3 | 7 | 206.5 | 50 | 205.0 ± 32.5 | |
| 4 | 8 | 236.0 | 80 | (6.95 ± 1.10%, v/v) | 0.995 |
| 5 | 9 | 265.5 | 100 | | |
| 6 | 10 | 295.0 | 100 | | |

## Chapter 8
## PULMONARY AND SYSTEMIC VASCULAR RESISTANCE IN THE CAT

From the review section (Chapter 6) it is apparent that the authors of this monograph are unaware of reports concerning circulatory responses during inhalation of MEK vapor. The present investigation was therefore undertaken to study the effects of graded concentrations of MEK on systemic and pulmonary circulatory dynamics in the cat. This section describes such experiments, as well as attempts to elucidate the mechanism of action of MEK-induced changes in systemic and pulmonary vascular resistances. In addition, the threshold concentrations for MEK effects are reported.

### A. Methods

Eight cats of either sex, weighing 2.5 to 3.6 kg (mean ± SE, 3.0 ± 0.2 kg), were anesthetized with pentobarbital sodium, 30 mg/kg, by intraperitoneal injection. The left femoral artery and vein were cannulated, and blood pressure recorded by a Statham P23AA transducer connected to a Sanborn 7700 recorder. After tracheal cannulation, mechanical ventilation was begun using a Harvard small animal respirator (20 breaths/min, 30 to 50 ml/breath). The thorax was then opened by an incision along the fifth left intercostal space. A polyethylene catheter (PE 50) was inserted into the small pulmonary artery leading to the lower left lobe, and connected to a Statham P23BB low pressure transducer for monitoring pulmonary arterial pressure. Another catheter (PE 60) was inserted into the left atrial appendage and advanced into the atrium for measurement of left atrial pressure (P23BB transducer). Pulmonary arterial blood flow, also representing cardiac output, was monitored by a Statham noncannulating flow probe (6 to 8 mm diameter, model SP 7515) placed around the main pulmonary artery and connected to a Statham SP 2202 flowmeter. Following these procedures, the animal was allowed to stabilize for 30 to 45 min, during which time the lungs were periodically expanded (sigh maneuver) to prevent atelectasis.

Vapor concentrations of MEK for inhalation studies were prepared by volatilizing a measured volume of MEK liquid into a stream of air entering a breathing bag at a set rate. The bag was connected to the inlet of the respirator during vapor inhalation. Following a control period of ventilation with room air, the cat was exposed to MEK vapors according to a sequence of increasing concentrations (v/v): 0.01% (0.3 mg/l), 0.05% (1.5 mg/l), 0.10% (3.0 mg/l), 0.25% (7.4 mg/l), 0.50% (14.8 mg/l), and 1.0% (29.5 mg/l). Exposures were carried out for 5 min each. In some experiments, the effects of various blocking agents on the response to MEK were tested following the dose-response procedures. After ventilating with room air and then 1.0% (29.5 mg/l) MEK for 5 min, the cat was given either the alpha-adrenergic blocking agent tolazoline hydrochloride (Priscoline®, Ciba-Geigy Pharmaceutical Co.), 2 to 10 mg/kg, i.v., or the antihistamine, diphenhydramine hydrochloride (Benadryl®, Parke, Davis and Co.), 2 to 10 mg/kg, i.v. When the animal had stabilized, a test inhalation of 1.0% MEK was then administered.

### B. Calculations and Abbreviations of Circulatory Measurements

MPAP: Mean pulmonary arterial pressure in cm $H_2O$, measured from the pulmonary artery to the left lower lobe.

MLAP: Mean left atrial pressure in cm $H_2O$, measured from the left atrium.

EMPAP: Effective mean pulmonary arterial pressure in cm $H_2O$, mean pulmonary arterial pressure minus mean left atrial pressure.

MAP: Mean arterial pressure in mm Hg, measured from the left femoral artery.

MPAF: Mean pulmonary arterial flow in ml/min, measured with a Statham electromagnetic flow probe around the main pulmonary artery, using a Statham electromagnetic flowmeter.

PVR: Pulmonary vascular resistance in dynes·sec/$cm^5$, the quotient of mean pulmonary arterial pressure minus mean left atrial pressure in dynes/$cm^2$ and of mean pulmonary arterial flow in ml/sec.

SVR: Systemic vascular resistance in dynes·sec/$cm^5$, the quotient of mean arterial pressure and of mean pulmonary arterial flow.

The paired t-test was used to determine the statistical significance of mean differences in values from control and experimental trials. Differences with P values of 0.05 and less were considered statistically significant.

**C. Results**

Portions of the record from a representative experiment consisting of inhalation of MEK are shown in Figure 8.1. The most pronounced and consistent effect of MEK on measured parameters is a concentration-dependent increase over controls in mean pulmonary arterial pressure (MPAP). Lesser increases can also be seen in systemic arterial pressure (AP) and mean pulmonary arterial flow (MPAF). In general, alterations in the last parameter were small and variable.

In Figure 8.2 are shown log concentration-response curves, expressing the major circulatory alterations in percent change from control averages. MEK effects on pulmonary and systemic circulatory factors, both measured and derived, are also summarized for eight experiments in Table 8.1. MEK was found to have a similar overall action, vasoconstriction, in both the pulmonary and systemic vascular beds.

*1. Pulmonary Vasoconstriction*

By far, the major effect of MEK is pulmonary hypertension accompanied by pulmonary vasoconstriction. Table 8.1 and Figure 8.2a show that while all concentrations were effective, the first significant ($P < 0.05$) increases in pulmonary vascular resistance (12%) and in mean pulmonary arterial pressure (8%; not shown in Figure 8.2) occurred following inhalation of 0.1% MEK. Effective mean pulmonary arterial pressure was also elevated (10%) but this was not statistically significant. Higher concentrations elicited further pulmonary hypertension and pulmonary vasoconstriction. In Figure 8.2a it can be seen that inhalation of 0.25% MEK caused elevations over controls of 19% for effective mean pulmonary arterial pressure and 23% for pulmonary vascular resistance, while 0.5% MEK elicited increases of 22% and 28%, respectively. Simultaneously, mean pulmonary arterial flow tended to decrease (3 to 5%), although this was not significant. Inhalation of 1.0% MEK abolished both the decrease in flow and any further elevation in derived pulmonary vascular resistance, although effective mean pulmonary arterial pressure continued to increase (approximately 30% over controls).

*2. Systemic Vasoconstriction*

As can be seen in Table 8.1 and Figure 8.2b, overall effects of MEK on the systemic circulation, including vasoconstriction, are much less than on the pulmonary circulation. Concentrations of 0.01% and 0.05% were without effect. A minimum effective concentration, 0.1%, slightly but significantly elevated mean arterial pressure (4%). Systemic vascular resistance also increased by 7% (nonsignificant), while mean pulmonary arterial flow, representing cardiac output, was unchanged (Figure 8.2a). Statistically significant increases over controls in calculated values for systemic resistance (7% and 11%) appeared during inhalation of 0.25% and 0.5% MEK, respectively. These changes were associated with a slight elevation in mean arterial pressure (3% and 4%) and decline in mean pulmonary arterial flow (3% and 5%). Inhalation of 1.0% MEK elicited no further increases in systemic arterial pressure or systemic vascular resistance. Rather, the recovery of blood flow to control levels, and the absence of any further increase over controls in mean arterial pressure (4%), led to modification (and loss of statistical significance) in the magnitude of elevated systemic vascular resistance (7%).

*3. Effect of Blocking Drugs on the Responses to MEK*

The alpha-adrenergic blocking drug, tolazoline, and the antihistaminic agent, diphenhydramine, were tested for possible effects in blocking the MEK-induced changes in pulmonary vascular resistance and other circulatory factors. A summary of the alterations in the MEK response following tolazoline in three cats is presented in Table 8.2. Overall, the largest MEK-induced increases, which occurred in factors affecting the pulmonary circulation, were reduced in magnitude to about 33 to 50% of the elevations found with MEK alone. As shown in Table 8.2, prior to alpha-adrenergic blockade, 1.0% MEK inhalation elicited significant elevation in mean pulmonary arterial pressure, effective mean pulmonary arterial pressure, and pulmonary vascular resistance (13%, 23%, and 20%, respectively). Tolazoline injection (2 to 10 mg/kg i.v.) substantially modified these increases (4%, 9%, and 10%, respectively), although they remained statistically significant. Part of this reduction may have been due to a tolazoline-induced increase (12 to 16%) in pulmonary vascular tone. Mean arterial pressure and systemic vascular resistance, which had been slightly elevated by MEK alone (4 and 3%, respectively), were now lower than controls (−6 and −4%, respectively).

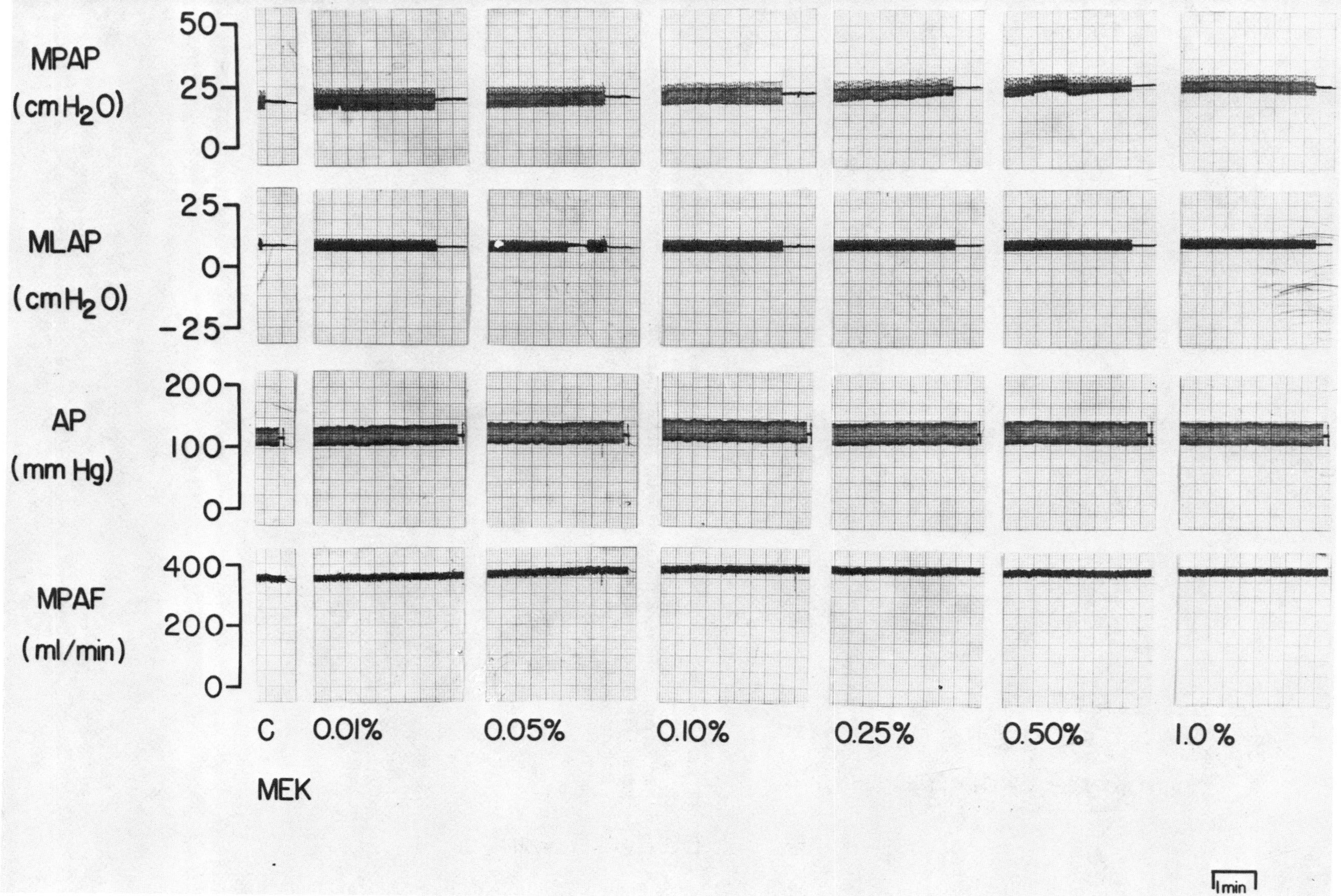


FIGURE 8.1. The effects of progressively increasing concentrations of MEK on some circulatory parameters in a representative experiment. The record shown represents the final 3 min of each 5-minute exposure. MPAP: mean pulmonary arterial pressure; MLAP: mean left atrial pressure; AP: arterial pressure; MPAF: mean pulmonary arterial flow.

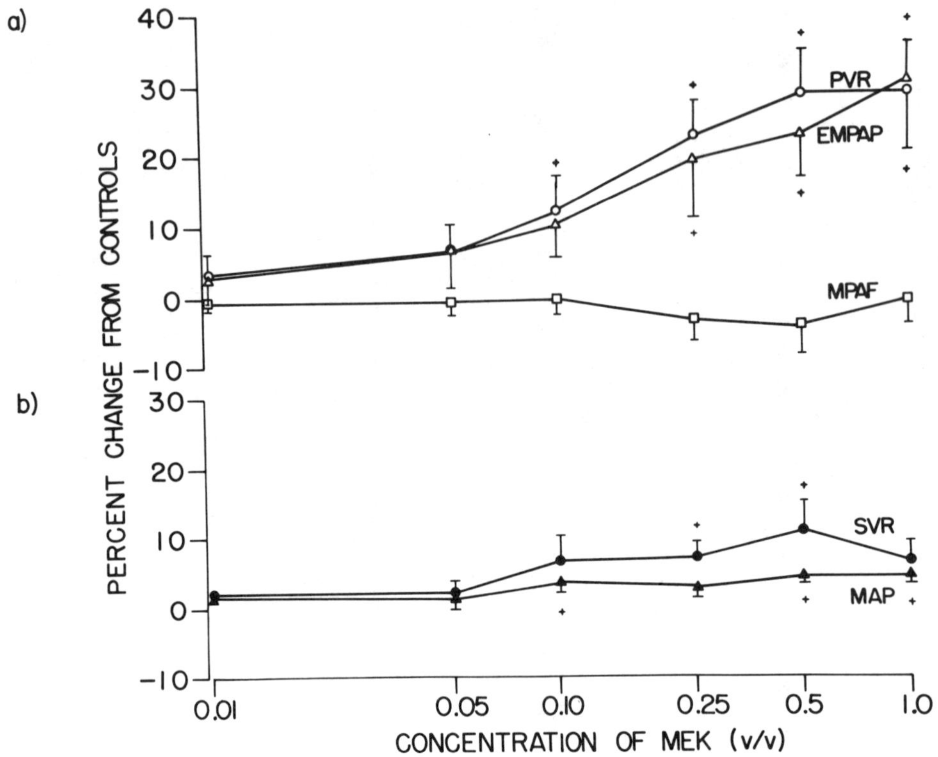


FIGURE 8.2. Mean percentage changes from control averages in response to progressively increasing concentrations of MEK in eight cats. (a) EMPAP: effective mean pulmonary arterial pressure; PVR: pulmonary vascular resistance; and MPAF: mean pulmonary arterial flow. (b) SVR: systemic vascular resistance; MAP: mean arterial pressure. Bars represent the SE of the mean. + indicates statistically significant mean differences between control and experimental trials, using the Student's t-test for paired data ($P \leqslant 0.05$).

TABLE 8.1

**Effect of Inhalation of Increasing Concentrations of MEK on Circulatory Dynamics in the Cat***

| | MPAP cm $H_2O$ | | MLAP cm $H_2O$ | | EMPAP cm $H_2O$ | | MAP mmHg | |
|---|---|---|---|---|---|---|---|---|
| | C | E | C | E | C | E | C | E |
| MEK 0.01% | 28.0 | 28.3 | 11.0 | 11.4 | 16.9 | 16.9 | 124.5 | 122.9 |
| | 0.4 ± 0.64 | | 0 ± 0.4 | | 2.6 ± 1.81 | | 2.0 ± 1.11 | |
| | NS | | NS | | NS | | NS | |
| MEK 0.05% | 28.0 | 29.5 | 11.0 | 11.5 | 16.9 | 18.0 | 124.5 | 126.0 |
| | 1.5 ± 0.80 | | 0.5 ± 0.57 | | 1.1 ± 0.77 | | 1.5 ± 1.49 | |
| | NS | | NS | | NS | | NS | |
| MEK 0.10% | 28.0 | 30.2 | 11.0 | 11.6 | 16.9 | 18.6 | 124.5 | 129.6 |
| | 2.2 ± 0.73 | | 0.6 ± 0.53 | | 1.7 ± 0.73 | | 5.1 ± 2.11 | |
| | 0.05 | | NS | | NS | | 0.05 | |
| MEK 0.25% | 28.0 | 31.4 | 11.0 | 11.7 | 16.9 | 19.8 | 124.5 | 128.6 |
| | 3.4 ± 1.11 | | 0.7 ± 0.61 | | 2.9 ± 1.08 | | 4.1 ± 1.89 | |
| | 0.02 | | NS | | 0.05 | | NS | |
| MEK 0.50% | 28.0 | 32.0 | 11.0 | 11.7 | 16.9 | 20.3 | 124.5 | 130.0 |
| | 4.0 ± 1.09 | | 0.6 ± 0.83 | | 3.4 ± 1.05 | | 5.5 ± 1.39 | |
| | 0.01 | | NS | | 0.02 | | 0.01 | |
| MEK 1.0% | 28.0 | 33.2 | 11.0 | 11.9 | 16.9 | 21.3 | 124.5 | 130.0 |
| | 5.2 ± 1.30 | | 0.9 ± 0.87 | | 4.5 ± 1.35 | | 5.5 ± 1.39 | |
| | 0.01 | | NS | | 0.02 | | 0.01 | |

*Data shown are mean control and experimental values, and the mean ± SE of the difference between these values, summarized from eight experiments. The t-test for paired data was used to determine statistical significance of the mean differences. MPAP: mean pulmonary arterial pressure; MLAP: mean left atrial pressure; EMPAP: effective mean pulmonary arterial pressure; MAP: mean arterial pressure; MPAF: mean pulmonary arterial flow.

TABLE 8.1 (continued)

**Effect of Inhalation of Increasing Concentrations of MEK on Circulatory Dynamics in the Cat***

| | MPAF ml/min | | Vascular resistances dynes•sec/$cm^5$ Pulmonary | | Vascular resistances dynes•sec/$cm^5$ Systemic | |
|---|---|---|---|---|---|---|
| | C | E | C | E | C | E |
| MEK 0.01% | 278.6 | 277.9 | 3529.6 | 3636.6 | 35372.4 | 36247.0 |
| | −0.7 ± 3.17 | | 107.0 ± 106.11 | | 874.6 ± 326.40 | |
| | NS | | NS | | NS | |
| MEK 0.05% | 278.6 | 279.3 | 3529.6 | 3712.6 | 35372.4 | 36502.9 |
| | 0.7 ± 5.39 | | 183.0 ± 142.89 | | 1130.4 ± 724.53 | |
| | NS | | NS | | NS | |
| MEK 0.10% | 278.6 | 273.6 | 3529.6 | 3940.3 | 35372.4 | 37898.9 |
| | −5.0 ± 7.15 | | 410.7 ± 159.17 | | 2530.7 ± 1186.26 | |
| | NS | | 0.05 | | NS | |
| MEK 0.25% | 278.6 | 271.4 | 3529.6 | 4268.0 | 35372.4 | 37929.9 |
| | −7.6 ± 8.79 | | 738.4 ± 187.63 | | 2557.4 ± 814.96 | |
| | NS | | 0.01 | | 0.05 | |
| MEK 0.50% | 278.6 | 267.1 | 3529.6 | 4479.7 | 35372.4 | 38943.9 |
| | −11.4 ± 11.69 | | 965.9 ± 244.06 | | 3571.4 ± 1100.1 | |
| | NS | | 0.01 | | 0.02 | |
| MEK 1.0% | 278.6 | 277.1 | 3529.6 | 4481.4 | 35372.4 | 37692.2 |
| | −1.4 ± 9.92 | | 951.4 ± 230.93 | | 2319.9 ± 1057.4 | |
| | NS | | 0.01 | | NS | |

Two additional experiments were carried out to determine a possible role for histamine release in the pulmonary vasoconstriction and hypertension elicited by inhalation of 1.0% MEK. In the first experiment, the record of which is presented in Figure 8.3, MEK caused increases in mean pulmonary vascular resistance (15%, 28%, and 26%, respectively). Following administration of diphenhydramine (2 mg/kg i.v.), MEK inhalation elicited a decrease from control averages in these parameters (−4%, −8%, and −5%, respectively). Similar, but less pronounced, reductions (by about 50%) in MEK effects were also found in the second experiment in which 10 mg/kg of diphenhydramine were administered.

### D. Discussion

These experiments demonstrate that in the cat, MEK is a potent pulmonary vasoconstrictor and hypertensive agent. Inhalation of as little as 0.01% (0.3 mg/l) MEK elicited slight nonsignificant elevations in both pulmonary vascular resistance (4%) and effective mean pulmonary arterial pressure (3%), 0.1% (3.0 mg/l) MEK, significant ($P < 0.05$) elevations in pulmonary vascular resistance (12%) and mean pulmonary arterial pressure (8%), and 1.0% (29.5 mg/l) MEK, maximal increases in pulmonary vascular resistance (29%) and effective mean pulmonary arterial pressure (30%). By contrast, systemic vascular resistance and mean arterial pressure were elevated only at higher concentrations (0.1% or more MEK) and to a much lesser extent, with maximal increases of 11% and 4%, respectively, occurring during inhalation of 0.5% (14.8 mg/l) MEK. An additional factor in the derived resistance values, a small, nonsignificant decline in mean pulmonary arterial flow (or cardiac output) first seen during inhalation of 0.25% (7.4 mg/l) MEK, appeared to contribute much more to the rises in systemic rather than pulmonary resistance. Thus, while recovery of flow during inhalation of 1.0% MEK obtunded any further rise in pulmonary vascular resistance, it substantially reduced the magnitude and abolished

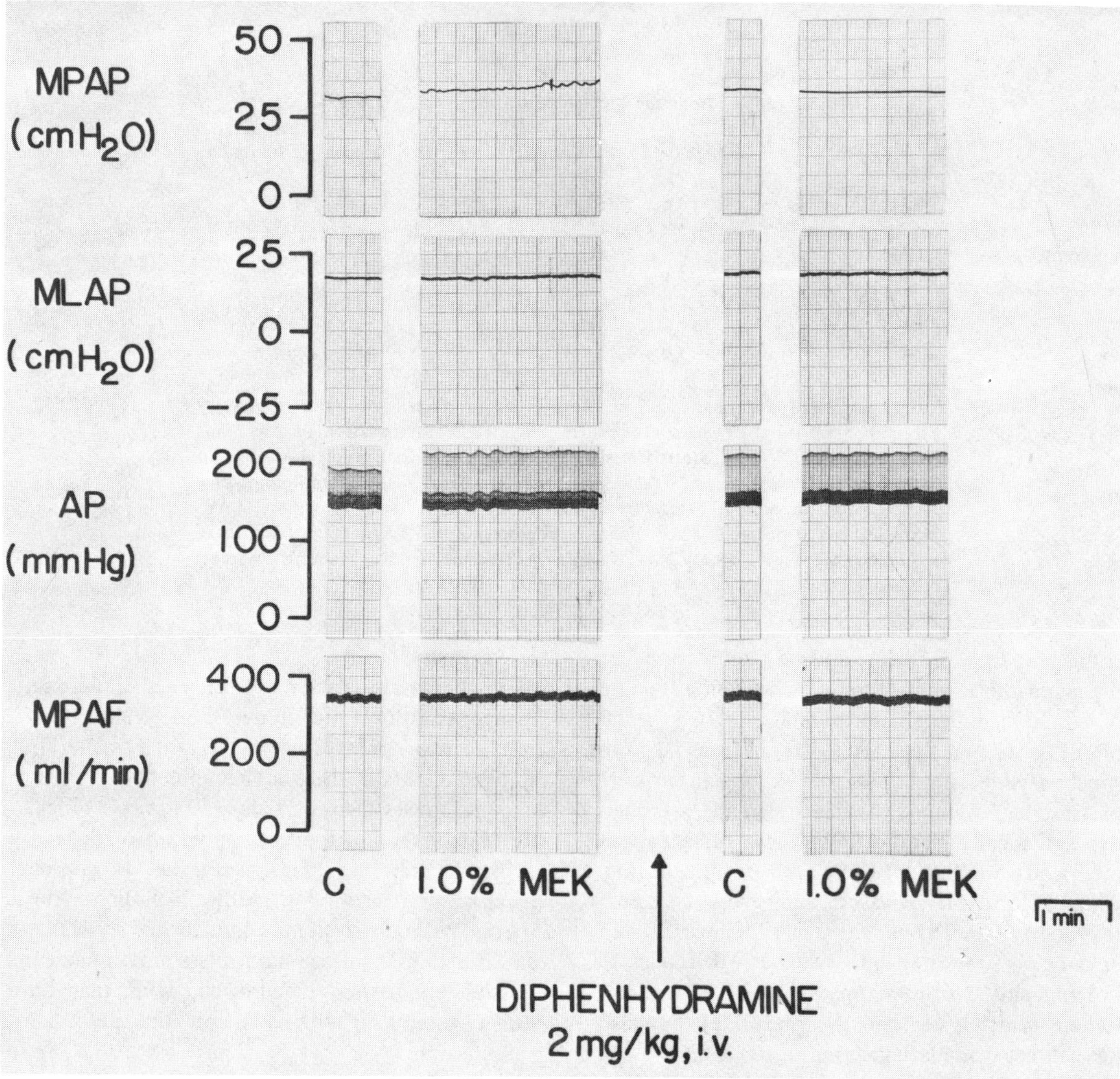


FIGURE 8.3. The effect of diphenhydramine injection (2 mg/kg, i.v.) on circulatory responses to inhalation of 1.0% MEK. The record shown represents the final portion of each 5-min exposure. MPAP: mean pulmonary arterial pressure; MLAP: mean left atrial pressure; AP: arterial pressure; MPAF: mean pulmonary arterial flow.

the statistical significance of the elevation in systemic vascular resistance.

The manner by which MEK elicits these changes, primarily in pulmonary factors, was next investigated by administering blocking drugs and then retesting the effects of administration of 1.0% MEK. Results with tolazoline, 2 to 10 mg/kg, i.v., indicate that alpha-adrenergic receptor stimulation could account for all of the rise in systemic factors induced by MEK, but only a part (50 to 66%) of the increase over controls in mean pulmonary arterial pressure and pulmonary vascular resistance, respectively (Table 8.2). The 50 to 100% effectiveness of diphenhydramine in abolishing the pulmonary circulatory response to MEK in two experiments suggests that histamine release, perhaps from pulmonary mast cells, may occur during MEK inhalation. This factor could explain the incomplete abolition of the pulmonary response during $\alpha$-adrenergic blockade.

The manner by which $\alpha$-adrenergic receptors may mediate the pulmonary vasoconstriction response to MEK is not fully explained on the basis of the current data. The $\alpha$-adrenergic stimu-

TABLE 8.2

**Effect of Alpha-adrenergic Blockade on Responses to 1.0% MEK**

| | % Change before tolazoline | % Change after tolazoline |
|---|---|---|
| MPAP | 12.6 ± 1.83* | 4.1 ± 0.26** |
| EMPAP | 23.2 ± 2.81** | 8.6 ± 1.72*** |
| MLAP | 0.7 ± 4.35 | −3.1 ± 1.66 |
| MAP | 4.1 ± 2.07 | −5.5 ± 3.37 |
| MPAF | 1.3 ± 3.25 | −1.8 ± 3.12 |
| PVR | 19.5 ± 4.60*** | 10.3 ± 3.33*** |
| SVR | 3.0 ± 4.44 | −3.8 ± 1.70 |

Values represent the mean ± SE percentage change from controls in circulatory parameters in three experiments, before and after administration of tolazoline HCl, 2 to 10 mg/kg, i.v. Statistical significance is determined using the t-test for paired data, *($P \leqslant 0.01$), **($P \leqslant 0.02$), ***($P \leqslant 0.05$). MPAP: mean pulmonary arterial pressure; EMPAP: effective mean pulmonary arterial pressure; MLAP: mean left atrial pressure; MAP: mean arterial pressure; MPAF: mean pulmonary arterial flow; PVR: pulmonary vascular resistance; SVR: systemic vascular resistance.

lating property of MEK could be from actions on the sympathetic nerve terminals, as well as release of adrenomedullary catecholamines. The reversal of MEK systemic effects by tolazoline, i.e., the small decline from controls in mean arterial pressure and systemic vascular resistance, although not statistically significant in these three experiments, suggests that MEK may release epinephrine, whose beta-adrenergic vasodilating property is now revealed (epinephrine reversal). However, an equally plausible possibility is that MEK releases norepinephrine from sympathetic nerve terminals (where effect is blocked by tolazoline), but also has a direct vasodilating action of its own.

## E. Summary

In the cat, inhalation of MEK elicits pulmonary vasoconstriction and hypertension (about 30% over control averages) as well as much smaller, but similar, effects in the systemic circulation. Pulmonary responses were partially (50 to 66%) blocked by tolazoline, and more variably (50 to 100%), by diphenhydramine. This suggests a complex mechanism of action of MEK, including sympathetic nervous system stimulation, as well as catecholamine release and histamine release as possible mediators. In addition, MEK may have direct effects of its own on the pulmonary vasculature.

# Chapter 9
# HEMODYNAMIC EFFECTS OF METHYL ETHYL KETONE INHALATION IN THE DOG

Methyl ethyl ketone is the first known solvent to cause pulmonary vasoconstriction in the cat lung (Chapter 8). Since the other solvents and propellants have only been tested in the dog lung, it is necessary to investigate the effects of MEK in the dog with hemodynamic measurements of the pulmonary and systemic circulations.

The inhalation of low oxygen and high carbon dioxide mixtures is known to produce pulmonary hypertension in the dog and cat and in man.[69] There is no reason to suspect that the occurrence of pulmonary vasoconstriction in one species precludes its occurrence in other animal species and in man. However, since solvents in general are not known to constrict the pulmonary blood vessels, it is important to establish that a particular solvent, i.e., MEK, can cause pulmonary hypertension in at least two species. If this is established, then the probability of its occurrence in man is increased so that workers exposed to MEK will have to be examined for signs of pulmonary hypertension.

## A. Methods

Seven adult mongrel dogs of either sex, weighing between 17 and 23 kg (average 19 kg), were anesthetized with 30 to 35 mg/kg pentobarbital sodium injected intravenously. The dogs were artificially ventilated with room air via endotracheal tubing and the aortic pressure was measured by inserting a catheter connected to a pressure transducer (Statham P23AA) into the right carotid artery. The chest was then opened on the left side at the fourth intercostal space. The lungs were gently retracted and the pericardium was opened. The pulmonary artery was dissected free from the aorta and an electromagnetic flow probe with an internal diameter of approximately 10% less than the pulmonary artery was fitted around the latter shortly before its bifurcation. The pulmonary arterial flow was thus monitored using a Statham SP2202 electromagnetic flowmeter. Left atrial pressure and pulmonary arterial pressure were measured by inserting two catheters, connected to Statham P23BB pressure transducers, in the appendage of the left atrium and in the pulmonary artery to the left lower lobe, respectively. To measure the left ventricular pressure the chest was opened at the sixth intercostal space of the right side and a catheter was inserted into the left ventricular lumen via the myocardial apex. The signal was fed to a differentiating circuit (8814A HP derivative computer) and the left ventricular dp/dt was recorded. All parameters were recorded on a Sanborn 7700 six-channel recorder.

The preparation was allowed to stabilize for a period of up to 30 min, after which the following concentrations of methyl ethyl ketone in air were administered via the inlet of the respirator for 5 min: 0.05%, 0.10%, 0.25%, 0.50%, and 1.0%. The subsequent concentration was administered to the animal immediately after the preceding one. Various concentrations of MEK were prepared by volatilizing the appropriate amount of the solvent into a known volume of air as previously described.[70]

The minimal effective concentration of MEK was administered to the dog preparation, alone or mixed with the threshold concentration of trichlorofluoromethane (FC 11), previously determined in a similar preparation.[71]

All data were analyzed by paired comparison. The Student's t-test was used to test for any significant difference between various compounds when administered separately or together. In both cases the criterion for significance is P less than or equal to 0.05.

## B. Calculations and Abbreviations of Hemodynamic Measurements

MPAP: Mean pulmonary arterial pressure in cm $H_2O$, measured from the pulmonary artery to the left lower lobe.

MLAP: Mean left atrial pressure in cm $H_2O$, measured from the left atrium.

EMPAP: Effective mean pulmonary arterial pressure in cm $H_2O$, mean pulmonary arterial pressure minus mean left atrial pressure.

LVP: Left ventricular pressure in mm Hg, measured from a catheter in the left ventricular cavity inserted through the cardiac apex.

LVEDP: Left ventricular end-diastolic pressure in mm Hg, measured from the left ventricular pressure.

dp/dt: Maximal rate of rise of left ventricular

pressure in mm Hg/sec derived from the left ventricular pressure with a derivative computer, type HP 8814A.

MAP: Mean aortic pressure in mm Hg, measured from a catheter inserted through the carotid artery.

MPAF: Mean pulmonary arterial flow in ml/min, measured with a Statham electromagnetic flow probe around the main pulmonary artery, using a Statham electromagnetic flowmeter.

HR: Heart rate, in beats/min, computed from aortic pressure waves.

Systemic vascular resistance: In dynes·sec/cm$^5$; the quotient of mean aortic pressure minus left ventricular end-diastolic pressure in dynes/cm$^2$ and of mean pulmonary arterial flow in ml/sec.

Pulmonary vascular resistance: In dynes·sec/cm$^5$; the quotient of mean pulmonary arterial pressure minus mean left atrial pressure in dynes/cm$^2$ and of mean pulmonary arterial flow in ml/sec.

Stroke volume: In ml, computed from cardiac output and heart rate.

Stroke work: In gram·meter (g·m) computed from stroke volume and aortic pressure.

**C. Results**

Methyl ethyl ketone inhaled in a concentration range of 0.05 to 1.0% in air was investigated in the open-chest dog preparation. The results of seven experiments are summarized in Table 9.1 and a record of a typical experiment is shown in Figure 9.1. In a concentration of 0.05%, MEK brought about no significant change in any of the parameters studied. However, increasing the concentration of the vapors of MEK in the inhaled mixture led to significant changes in various parameters as follows.

*1. Pulmonary Arterial Pressure and Pulmonary Vascular Resistance*

In as low a concentration as 0.10%, MEK brought about a significant increase of 2.6% in pulmonary arterial pressure, 3.6% in effective mean pulmonary arterial pressure, and 4.6% in pulmonary vascular resistance. A maximum increase of 4.8%, 8.7%, and 11.6% in these three parameters, respectively, was achieved by inhalation of 1.0% of the ketone. The consistent, significant, and concentration-dependent increase in pulmonary vascular resistance is depicted in Figure 9.2. Moreover, a nonsignificant increase in systemic vascular resistance was achieved by various concentrations of MEK.

*2. Systemic Arterial Pressure*

A significant increase of 4.3% in mean arterial pressure as compared with the control value was elicited by 0.1% MEK. This rise in blood pressure declined with the increase in the concentration inhaled. A concomitant increase in heart rate averaging 2.5% over the control value and a decrease in myocardial contractility of 3.4% was elicited by inhaling 0.1% of the ketone.

*3. Heart Rate*

Methyl ethyl ketone brought about a positive chronotropic effect in all concentrations used, especially with 0.50% where the increase in heart rate reached a maximum of 4.6% as compared with the control value. In spite of this increase in heart rate, the cardiac output was decreased by 2.5% with the same concentration of MEK, viz., 0.50%. Nonetheless, the decrease in cardiac output elicited by various concentrations of the ketone (0.1 to 1.0%) is still nonsignificant.

*4. Stroke Volume and Work*

The inhalation of 0.5% v/v of methyl ethyl ketone brought about a significant decrease in stroke volume (0.6 ± 0.17 ml) and stroke work (1.2 ± 0.35 g·m). The mean decrease in cardiac output achieved by the same concentration of the ketone is 33.3 ± 16.7 ml/min though this is not a significant change.

*5. Interaction of MEK and Other Solvents and Propellants*

Because of the likelihood of using methyl ethyl ketone mixed with a propellant in consumer products, it was found necessary to study any possible interaction between this ketone and trichlorofluoromethane (FC 11). For this reason, the minimal effective concentration of MEK was determined in various dog preparations and ranged from 0.5% to 1.0% with a mean concentration of 0.09 ± 0.01%. The various cardiovascular effects of this threshold concentration alone or mixed with the threshold concentrations of FC 11 were studied. Results are summarized in Table 9.2. At the concentration of 0.09% MEK elicited an increase in mean pulmonary arterial pressure, effective mean pulmonary arterial pressure, left

TABLE 9.1

**Myocardial and Hemodynamic Effects of Various Concentrations of Methyl Ethyl Ketone (MEK) on the Open-chest Dog Preparation***

| | MPAP cm $H_2O$ | MLAP cm $H_2O$ | EMPAP cm $H_2O$ | LVP mmHg | LVEDP mmHg | dp/dt mmHg/sec | MAP mmHg | MPAF ml/min | HR beats/min | Vascular resistance dynes·sec/cm$^5$ Pulmonary | Vascular resistance dynes·sec/cm$^5$ Systemic | Stroke vol ml | Stroke work g·m |
|---|---|---|---|---|---|---|---|---|---|---|---|---|---|
| Control | 33.7 ± 0.88 | 5.1 ± 0.68 | 28.6 ± 1.42 | 138.5 ± 6.69 | 2.3 ± 0.79 | 4113 ± 203.7 | 130.1 ± 4.04 | 1514 ± 122.5 | 167 ± 5.1 | 1150 ± 57.9 | 7111 ± 546.4 | 9.2 ± 0.81 | 17.3 ± 1.58 |
| 0.05% MEK | 33.3 ± 0.95<br>+0.8 ± 0.41<br>NS | 5.5 ± 0.77<br>+0.3 ± 0.35<br>NS | 27.9 ± 1.58<br>+0.5 ± 0.35<br>NS | 136.8 ± 11.05<br>+0.7 ± 1.09<br>NS | 2.5 ± 1.12<br>+0.8 ± 0.83<br>NS | 4250 ± 282.3<br>+25 ± 72.9<br>NS | 128.2 ± 7.39<br>+3.0 ± 1.51<br>NS | 1567 ± 203.2<br>+16.7 ± 21.1<br>NS | 164 ± 10.1<br>+2.6 ± 1.1<br>NS | 1103 ± 99.7<br>+4.5 ± 18.3<br>NS | 6855 ± 853.5<br>+67 ± 115.3<br>NS | 9.9 ± 1.45<br>−0.1 ± 0.15<br>NS | 18.3 ± 2.94<br>+0.2 ± 0.42<br>NS |
| 0.10% MEK | 33.3 ± 0.92<br>+0.8 ± 0.31<br>0.05 | 5.1 ± 0.81<br>−0.1 ± 0.35<br>NS | 28.3 ± 1.58<br>+0.9 ± 0.28<br>0.05 | 137.7 ± 9.53<br>+1.5 ± 2.14<br>NS | 3.8 ± 1.80<br>+2.1 ± 1.00<br>NS | 4050 ± 251.9<br>−175 ± 192.0<br>NS | 130.2 ± 5.78<br>+5.0 ± 1.71<br>0.05 | 1533 ± 199.03<br>−16.7 ± 21.8<br>NS | 165 ± 10.5<br>+4.2 ± 1.7<br>NS | 1142 ± 98.1<br>+42.6 ± 14.2<br>0.05 | 7040 ± 799.2<br>+252.8 ± 128.4<br>NS | 9.7 ± 3.13<br>−0.3 ± 0.17<br>NS | 18.3 ± 2.70<br>+0.1 ± 0.51<br>NS |
| 0.25% MEK | 33.7 ± 1.45<br>+0.7 ± 0.45<br>NS | 4.6 ± 1.16<br>−0.9 ± 0.36<br>NS | 28.2 ± 1.94<br>+0.7 ± 1.08<br>NS | 134.3 ± 10.61<br>−3.8 ± 1.83<br>NS | 4.2 ± 1.39<br>+2.5 ± 1.29<br>NS | 3925 ± 286.7<br>−300 ± 196.1<br>NS | 127.5 ± 6.92<br>+0.3 ± 0.33<br>NS | 1500 ± 204.9<br>−50 ± 22.4<br>NS | 163 ± 9.3<br>+2.2 ± 1.2<br>NS | 1169 ± 9.3<br>+63.0 ± 1.2<br>0.05 | 7159 ± 1058<br>+208.2 ± 202.3<br>NS | 9.5 ± 1.46<br>−0.5 ± 0.20<br>NS | 17.5 ± 2.82<br>−0.9 ± 0.46<br>NS |
| 0.50% MEK | 33.3 ± 1.03<br>+0.7 ± 0.38<br>NS | 5.1 ± 0.90<br>−0.3 ± 0.25<br>NS | 28.2 ± 1.64<br>+1.02 ± 0.62<br>NS | 137.7 ± 12.50<br>−0.5 ± 2.57<br>NS | 5.0 ± 1.71<br>+2.5 ± 0.91<br>0.05 | 4125 ± 285.0<br>−125 ± 147.9<br>NS | 127.2 ± 6.76<br>0 ± 0.52<br>NS | 1492 ± 211.5<br>−33.3 ± 16.7<br>NS | 168 ± 10.3<br>+7.6 ± 2.1<br>0.05 | 1183 ± 109.6<br>+71.0 ± 24.1<br>0.05 | 7129 ± 1017<br>+122 ± 185.1<br>NS | 9.2 ± 1.57<br>−0.6 ± 0.17<br>0.05 | 17.0 ± 3.01<br>−1.2 ± 0.35<br>0.05 |
| 1.0% MEK | 34.6 ± 1.43<br>+1.6 ± 0.59<br>0.05 | 4.4 ± 1.25<br>−0.7 ± 0.49<br>NS | 30.2 ± 1.93<br>+2.3 ± 0.61<br>0.02 | 145.7 ± 12.39<br>+3.7 ± 2.01<br>NS | 4.2 ± 2.01<br>+2.5 ± 1.12<br>NS | 4480 ± 3402<br>−20 ± 107.9<br>NS | 129.3 ± 7.93<br>+1.3 ± 1.20<br>NS | 1508 ± 224.5<br>−25 ± 40.3<br>NS | 167 ± 9.1<br>6.2 ± 2.9<br>NS | 1265 ± 127.8<br>+131.8 ± 39.3<br>0.02 | 7241 ± 1034<br>+184 ± 291.1<br>NS | 9.3 ± 1.47<br>−0.6 ± 0.40<br>NS | 17.6 ± 3.12<br>−0.7 ± 0.69<br>NS |

*Each set of numbers is pooled from 5–7 dogs and represents mean control ± SE, mean response ± S.E, mean difference ± SE of the difference, and the P value.

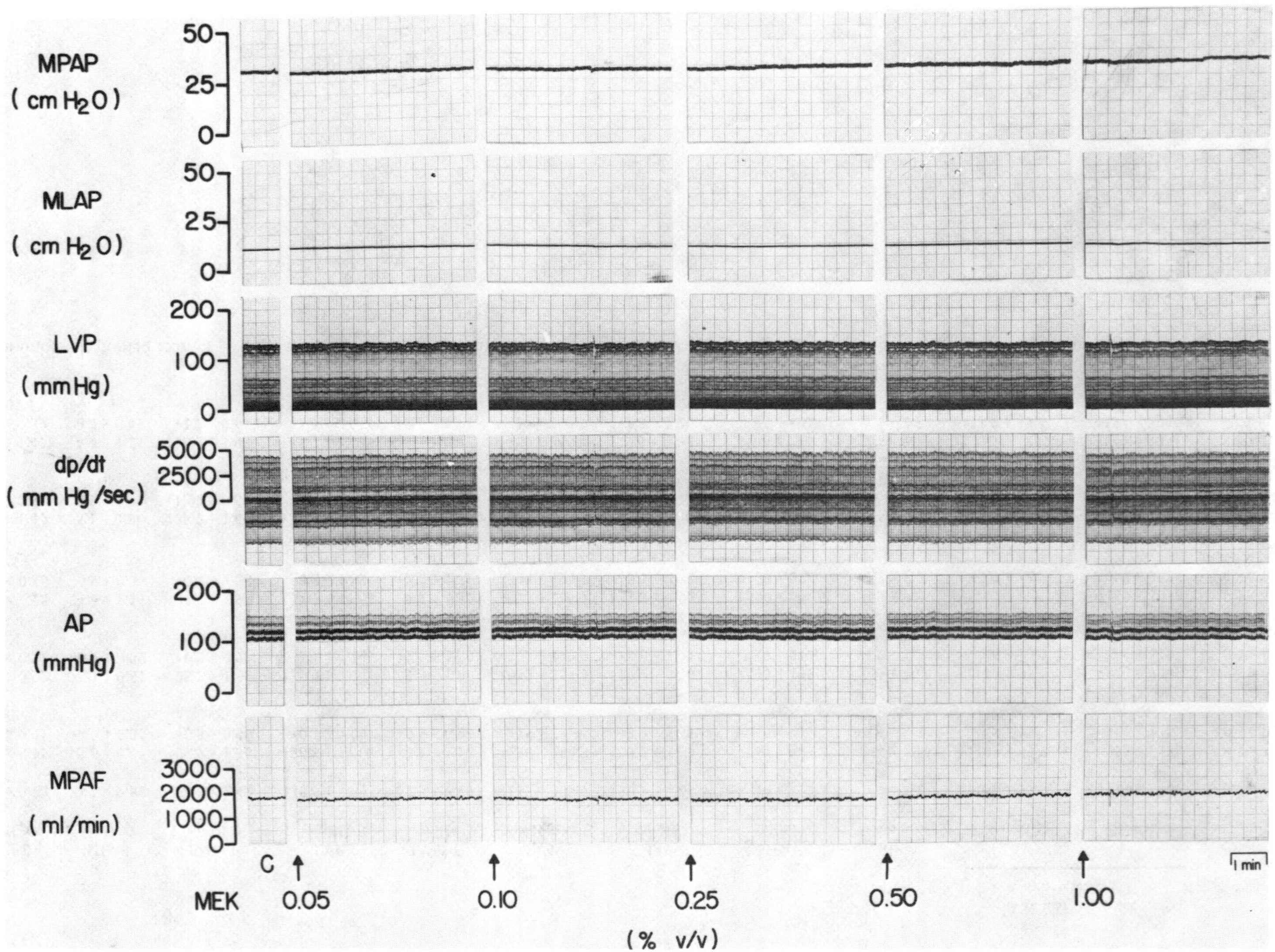


FIGURE 9.1. The effect of various progressively increasing (0.05% to 1.0%) concentrations of MEK on the myocardial contractility and hemodynamics of the open-chest dog preparation. Abbreviations are explained in the text.

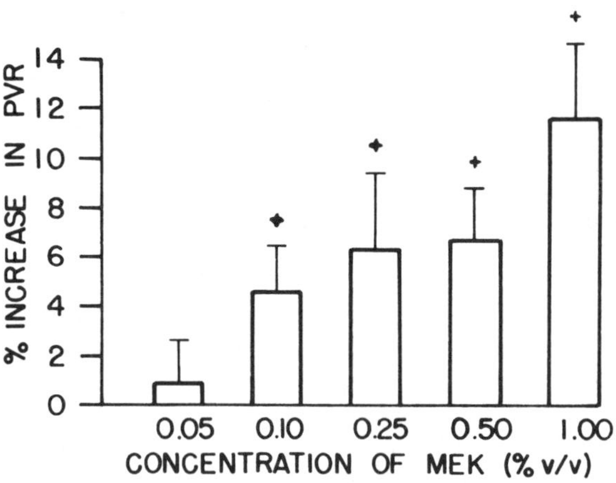


FIGURE 9.2. The effect of various concentrations of MEK on the pulmonary vascular resistance (PVR) in the open-chest dog preparation. + denotes significant change from the control value ($P < 0.05$).

ventricular dp/dt, mean aortic pressure, and pulmonary and systemic vascular resistances. Of special interest is the effect on the parameters depicted in Figure 9.3.

The minimal effective concentration of FC 11 was previously determined in a similar preparation to be 0.5%. This concentration brought about a decrease in left ventricular pressure, left ventricular dp/dt, mean aortic pressure, pulmonary and systemic vascular resistance, stroke volume, and stroke work (Figure 9.3 and Table 9.2). Administration of a mixture of the threshold concentrations of both MEK and FC 11, however, resulted in a decrease in left ventricular pressure, dp/dt, mean aortic pressure, stroke volume, and stroke work as shown in Figure 9.3 and Table 9.2. A significant difference was observed between the effect of MEK alone and that of a mixture of FC 11 and MEK on left ventricular pressure and dp/dt. However, the effect of the mixture on these two parameters does not significantly differ from that of FC 11 alone. This means that the effect of FC 11 predominates over that of the ketone on these previously mentioned parameters. On the other hand, a significant difference between the effect of FC 11 alone and that of the mixture of FC 11 and the ketone was observed on mean aortic pressure. Thus, while FC 11 decreased mean aortic pressure by 3.9%, yet a mixture of FC 11 and MEK brought about a decrease of about 1% in the same parameter. Thus the MEK effect predominates in this instance, though the net result is qualitatively as that of FC 11 alone.

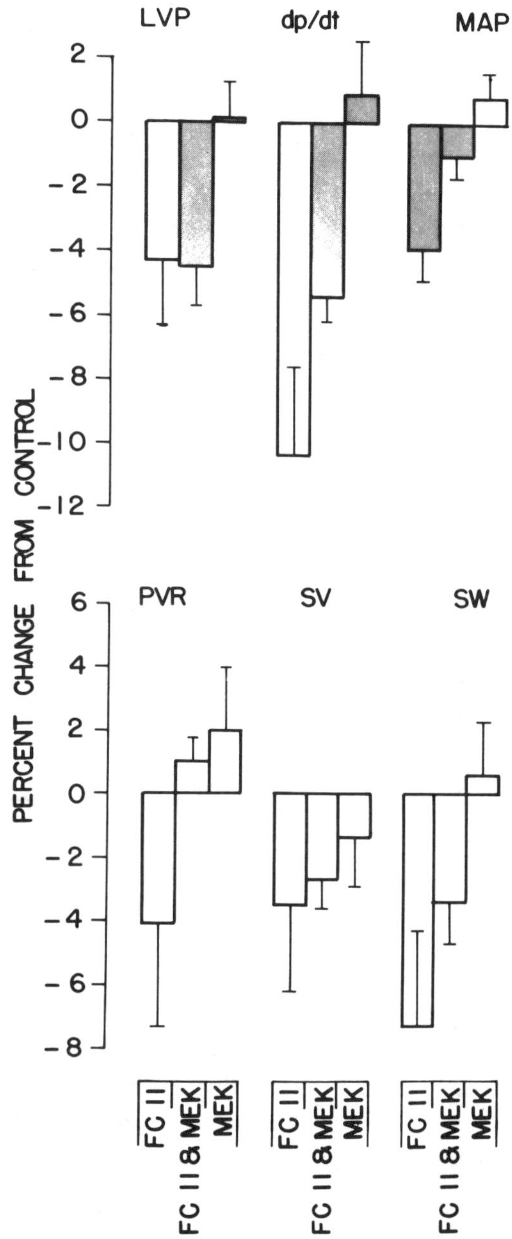


FIGURE 9.3. The effect of the threshold concentration of MEK (0.09%), threshold concentration (0.5%) of trichlorofluoromethane (FC 11), and a mixture of both on some of the measured (upper) and derived (lower) parameters of the open-chest dog preparation. Significant differences between two neighboring columns is indicated by hatched bars.

## D. Discussion

Methyl ethyl ketone, unlike isopropanol (Chapter 4), is effective in as low a concentration as 0.10%. The hemodynamic study of this com-

TABLE 9.2

**Myocardial and Hemodynamic Effects of the Threshold Concentration of MEK (0.09 ± 0.01%), Alone or Mixed with the Threshold Concentration of Trichlorofluoromethane (FC 11), 0.5%***

| | MPAP cm $H_2O$ | | MLAP cm $H_2O$ | | EMPAP cm $H_2O$ | | LVP mmHg | | LVEDP mmHg | | dp/dt mmHg/sec | |
|---|---|---|---|---|---|---|---|---|---|---|---|---|
| | C | E | C | E | C | E | C | E | C | E | C | E |
| MEK 0.09 ± 0.01% | 32.3 | 32.6 | 5.2 | 5.5 | 27.0 | 27.1 | 161 | 161 | 3.1 | 3.8 | 4531 | 4563 |
| | +0.32 ± 0.29 | | +0.3 ± 0.26 | | +0.1 ± 0.48 | | 0 ± 2.04 | | 0.63 ± 0.63 | | +32 ± 78.64 | |
| | NS | | NS | | NS | | NS | | NS | | NS | |
| FC 11 0.5% | 31.1 | 29.4 | 5.3 | 5.2 | 25.8 | 24.2 | 150 | 143 | 3.8 | 5.6 | 4156 | 3813 |
| | −1.8 ± 1.35 | | −0.2 ± 0.58 | | −1.6 ± 1.10 | | −7.0 ± 3.5 | | +1.8 ± 1.20 | | −343 ± 93.75 | |
| | NS | | NS | | NS | | NS | | NS | | 0.02 | |
| MEK (threshold) +0.05% FC 11 | 28.9 | 28.8 | 5.6 | 5.3 | 23.3 | 23.4 | 156 | 149 | 4.4 | 4.4 | 4031 | 3813 |
| | −0.1 ± 0.41 | | −0.3 ± 0.54 | | +0.1 ± 0.45 | | −7.0 ± 2.29 | | 0 ± 0 | | −218 ± 31.75 | |
| | NS | | NS | | NS | | 0.05 | | NS | | 0.01 | |

| | MAP mmHg | | MPAF ml/min | | HR beats/min | | Vascular resistance dynes·sec/$cm^5$ Pulmonary | | Vascular resistance dynes·sec/$cm^5$ Systemic | | Stroke vol ml | | Stroke work g·m | |
|---|---|---|---|---|---|---|---|---|---|---|---|---|---|---|
| | C | E | C | E | C | E | C | E | C | E | C | E | C | E |
| MEK 0.09 ± 0.01% | 139 | 140 | 1420 | 1410 | 173 | 174 | 1193 | 1204 | 8167 | 8332 | 8.3 | 8.2 | 16.1 | 16.3 |
| | +1.0 ± 1.0 | | −10 ± 18.71 | | +0.4 ± 0.93 | | +11 ± 21.44 | | +165 ± 183.38 | | −0.08 ± 0.10 | | +0.2 ± 0.33 | |
| | NS | | NS | | NS | | NS | | NS | | NS | | NS | |
| FC 11 0.5% | 142 | 136 | 1440 | 1390 | 170 | 170 | 1067 | 1024 | 8031 | 7810 | 8.6 | 8.3 | 17.3 | 16.1 |
| | −6.0 ± 1.60 | | −50 ± 38.73 | | +0.2 ± 1.66 | | −43 ± 36.71 | | −221 ± 298.35 | | −0.3 ± 0.26 | | −1.2 ± 0.58 | |
| | 0.05 | | NS | | NS | | NS | | NS | | NS | | NS | |
| MEK (threshold) +0.05% FC 11 | 139 | 138 | 1380 | 1370 | 169 | 172 | 1007 | 1019 | 8362 | 8374 | 8.2 | 8.0 | 16.3 | 15.8 |
| | −1.0 ± 0.98 | | −10 ± 24.5 | | +3.0 ± 2.16 | | +12.6 ± 10.33 | | +12 ± 190.77 | | −0.2 ± 0.06 | | −0.5 ± 0.15 | |
| | NS | | NS | | NS | | NS | | NS | | 0.05 | | 0.05 | |

*Each set of numbers is pooled from 5 dogs and represents mean control, mean experimental, mean difference ± SE of the difference, and the P value.

pound in the open-chest dog preparation revealed that the most consistent effect, in fact an effect which is achieved with 0.10% of the ketone, is an increase in the pulmonary vascular resistance. Increase of the concentration in the inhalation mixture to 0.5% brought about an increase in mean aortic pressure and heart rate and a decrease in stroke volume and stroke work.

*1. Effect on the Pulmonary Vascular Bed*

A consistent and significant increase in pulmonary vascular resistance was obtained by concentrations ranging from 0.1 to 1.0% of the ketone. Since the pulmonary vascular resistance is a derived parameter obtained by dividing the effective mean pulmonary arterial pressure by the pulmonary arterial flow, it reflects the relative changes in these two parameters and not necessarily the actual state of the pulmonary vascular bed. However, since decrease in pulmonary arterial flow brought about by 0.1% of the ketone is nonsignificant (0.8 ± 1.10%), comparison of the changes in pulmonary arterial pressure and aortic pressure might be helpful. The same concentration of methyl ethyl ketone, viz., 0.1%, brought about an increase in pulmonary arterial and aortic pressures of 2.6% and 4.3%, respectively. A decrease in cardiac output and pulmonary arterial flow tends to increase systemic and pulmonary vascular resistances. However, the increase in pulmonary arterial and aortic pressures (2.6% and 4.3%) is more than the decrease in cardiac output (0.8%). This suggests that the ketone brings about an increase in the vascular tone of these beds. The site of action and the mechanism of action of MEK, however, waits to be revealed by further investigation.

*2. Effect on Systemic Pressure*

It is known that MEK brings about, especially in high concentrations, a depression of the central nervous system. This central depressant effect tends to decrease the vascular tone and consequently decrease the blood pressure. However, results of these experiments revealed a significant increase in the aortic pressure averaging 4.3% of the control value at 0.1% concentration. This increase in aortic pressure decreased gradually with increase in the concentration of the inhaled ketone, reaching a level of 0.1% with 0.5% v/v of the ketone. Since this concentration (0.1%) of the ketone brought about a decrease in myocardial contractility of 3.4%, the significant increase in aortic pressure could be due to an increase in systemic resistance, which in turn is affected by several factors. The gradual decline in the increase of the aortic pressure with higher concentration is not difficult to appreciate, taking into account the central depressant effect of MEK. Gradual deepening of the central depressant effect tends to decrease the vascular tone which combats the increase in resistance brought about by different mechanisms.

*3. Effect on the Heart Rate*

An increase in aortic blood pressure is expected to bring about a reflex decrease in the heart rate through the pressoreceptors of the aortic arch and carotid sinus. However, MEK elicited a minor increase in both the aortic pressure and the heart rate. This tachycardia is gradually intensified reaching a maximum of 4.6% with 0.5% v/v of the ketone. This effect could be possibly achieved either by direct effect on the sinoatrial node, direct stimulation of beta-receptors, or release of catecholamines from their storage sites.

*4. Effect on Stroke Volume and Work*

The significant decrease in stroke volume brought about by MEK is expected from the decrease in cardiac output and increase in heart rate. The decrease in stroke volume might be due to the decrease in contractility and/or decrease in the effective filling pressure of the ventricles. The decrease in stroke work at 0.5%, despite the absence of stimulating changes in aortic pressure, can be explained by the decrease in stroke volume due to the significant increase in heart rate.

*5. Effect of Methyl Ethyl Ketone and FC 11*

Examination of Figure 8.3, which represents the effect of threshold concentrations of MEK, FC 11, and a mixture of both, revealed that at this concentration, MEK brought about a nonsignificant increase in left ventricular dp/dt and aortic pressure. On the other hand, FC 11 brought about a significant decrease in the same parameters. The effect of the mixture of threshold concentrations of MEK and FC 11 is qualitatively similar and not significantly different from those of FC 11 alone. This indicates that when FC 11 and MEK are mixed together, the effects of FC 11 dominate those of the ketone on the left ventricular pressure and dp/dt, as shown from the significant difference between their effects (hatched bars in Figure 9.3).

**E. Summary**

In the anesthetized open-chest dog preparation, the threshold effective concentration of MEK is 0.09 ± 0.01%. Inhalation of 0.1 to 1.0% v/v resulted in the following myocardial and hemodynamic changes: (a) increase in pulmonary arterial pressure and pulmonary vascular resistance; (b) increase in aortic pressure that gradually became less pronounced by increase in concentration due most probably to central and peripheral effects; (c) increase in heart rate suggesting an effect on the sinoatrial node, beta-receptors or release of catecholamines; and (d) decrease in stroke volume and stroke work especially with higher concentrations (0.5%).

When MEK was inhaled in the same dog preparation together with FC 11, the effect of the latter compound predominated over that of MEK on left ventricular pressure and dp/dt.

## REFERENCES – Methyl Ethyl Ketone


1. **Langedijk, S. L.**, Notes on methyl ethyl ketone (butanone), *Chem. Ind.* (London), 891, 1938.
2. **Browning, E.**, *Toxicity of Industrial Organic Solvents,* Her Majesty's Stationery Office, London, 1953.
3. **Smith, A. R. and Mayers, M. R.**, Poisoning and fire hazards of butanone and acetone, *Ind. Bull. N.Y. State Dep. Labor,* 23, 174, 1944.
4. **Press, E. and Done, A. K.**, Solvent sniffing: physiologic effects and community control measures for intoxication from the intentional inhalation of organic solvents, I., *Pediatrics,* 39, 451, 1967.
5. **Patty, F. A., Schrenk, H. H., and Yant, W. P.**, Acute response of guinea pigs to vapors of some new commercial organic compounds. VIII. Butanone, *U.S. Public Health Rep.,* 50, 1217, 1935.
6. **Gleason, M. N., Gosselin, R. E., Hodge, H. C., and Smith, R. P.**, *Clinical Toxicology of Commercial Products,* 3rd ed., Williams and Wilkins, Baltimore, 1969.
7. **Nelson, K. W., Ege, J. F., Jr., Morwick, R., Woodman, L. E., and Silverman, L.**, Sensory response to certain industrial solvent vapors, *J. Ind. Hyg. Toxicol.,* 25, 282, 1943.
8. **Jacobziner, H.**, Accidental chemical poisonings: glue sniffing, *N.Y. State J. Med.,* 63, 2415, 1963.
9. **Jacobziner, H.**, Accidental chemical poisonings, lead poisoning and glue sniffing intoxications, *N.Y. State J. Med.,* 63, 2846, 1963.
10. **Wilson, R. H., Hough, G. V., and McCormick, W. E.**, Medical problems encountered in the manufacture of American-made rubber, *Ind. Med.,* 17, 199, 1948.
11. **Munies, R. and Wurster, D. E.**, Investigation of some factors influencing percutaneous absorption. III. Absorption of methyl ethyl ketone, *J. Pharm. Sci.,* 54, 1281, 1965.
12. **Buell, G. C., Tokiwa, Y., Mueller, P. K.**, Potential crosslinking agents in lung tissue, *Arch. Environ. Health,* 10, 213, 1965.
13. **Smith, A. F. and Wood, R.**, A field test for the determination of some ketone vapours in air, *Analyst* (London), 97, 363, 1972.
14. **Kohli, R. P., Kishor, K., Dua, P. R., and Saxena, R. C.**, Anticonvulsant activity of some carbonyl containing compounds, *Indian J. Med. Res.,* 55, 11, 1967.
15. **Gautier, J. A., Farnoux, C. C., Nakajima, H., and Thuillier. J..** Activite anesthesique locale d'une serie d'amino-alcools et amino-cetones aromatiques a chaine butylique; etude des relations entre la structure et l'activite de ces composes, *Med. Pharmacol. Exp.,* 13, 325, 1965.
16. **Ringler, W. H.**, Two-step solvent dewaxing process (Standard Oil Co., Ohio). U.S. Patent 3,318,800, 1967.
17. **Woodle, R. A.**, High-grade lubricants from crude oil distillates (Texaco Development Corp.). German Patent 2,016,894, 1970.
18. Solid synthetic paper, (Sumitomo Chemical Co., Ltd.). French Application 2,000,883, 1969.
19. **Jensch, W.**, Elimination of surface adhesion of plasticized poly(vinyl butyral) sheets, (Dynamit Nobel A. G.). French Patent 1,598,247, 1970.
20. **Wippler, C.**, Cross-linking in poly(vinyl chloride) under the influence of ionizing radiation, (St. Gobain Factory, Antony, France), *J. Polym. Sci.,* 29, 585, 1958, in French, English summary.
21. **Ealding, C. J.**, Foamed thermoplastic polymers, (Imperial Chemical Industries, Ltd.). British Patent 1,220,053, 1971.
22. **Horton, R. A.**, Precision casting molds, (Precision Metalsmiths, Inc.). German Patent 1,223,502, 1966.
23. **Williams, H. L. and Davidson, M. J. G.**, Cleaning of polymerization reactors (Polymer Corporation, Ltd.). Canadian Patent 527,001, 1956.
24. American Conference of Governmental Industrial Hygienists, Documentation of the Threshold Limit Values for Substances in Workroom Air, 3rd ed., Cincinnati, Ohio, 1971.
25. **Berg, E. F.**, Retrobulbar neuritis, a case report of presumed solvent toxicity, *Ann. Opthalmol.,* 3, 1351, 1971.
26. **Osborne, J. S., Adamek, S., and Hobbs, M. E.**, Some components of gas phase of cigarette smoke, *Anal. Chem.,* 28, 211, 1956.
27. **Nefedov, Y. G., Savina, V. P., Sokolov, N. L., Ryzhkova, V. E.**, Contaminants in the air exhaled by man, *Kosm. Biol. Aviakosmicheskaya Med.,* 3, 71, 1969. (Russian).
28. **Tsao, M. U. and Pfeiffer, E. L.**, Isolation and identification of a new ketone body in normal human urine, *Proc. Soc. Exp. Biol. Med.,* 94, 628, 1957.
29. **Mabuchi, H.**, Urinary and serum volatile substances in diabetics, *Nippon Naika Gakkai Zasshi,* 58, 731, 1969 (Japanese).
30. **Wong, M. P. and Patton, S.**, Identification of some volatile compounds related to the flavor of milk and cream, *J. Dairy Sci.,* 45, 724, 1962.
31. **Harvey, R. J. and Walker, J. R. L.**, Some volatile compounds in New Zealand cheddar cheese and their possible significance in flavor formation. III. Time of first appearance of volatile carbonyl compounds during ripening, *J. Dairy Res.,* 27, 335, 1960.
32. **Schormueller, J. and Grosch, W.**, Aromatics of foods. II. Occurrence of additional carbonyl compounds in the tomato, *Z. Lebensm. Unters. Forsch.,* 126, 38, 1964.

33. **Dornseifer, T. P., Kim, S. C., Keith, E. S., and Powers, J. J.,** Effect of moisture level on volatile carbonyls in cottonseed oil heated to 210°, *J. Am. Oil Chem. Soc.*, 42, 1073, 1965.
34. **Shimizu, Y., Matsuto, S., Ito, Y., and Okada, I.,** Flavor of roast barley. III. Separation and identification of the volatile monocarbonyl compounds, *Nippon Nogei Kagaku Kaishi,* 43, 217, 1969.
35. **Bryce, D. J. and Greenwood, C. T.,** Physicochemical studies on starches. XXIX. Thermal degradation of starch. 3. Formation of decomposition products from starch and related materials at temperatures between 175° and 400°, *Staerke,* 15, 359, 1963.
36. **Cremer, E. and Riedmann, M.,** Identification of the gas-chromatographically separated aromatic materials of honey, *Z. Naturforsch.*, 19b, 76, 1964.
37. **Ng, H., Reed, D. J., and Pence, J. W.,** Identification of carbonyl compounds in an ethanol extract of fresh white bread, *Cereal Chem.*, 37, 639, 1960.
38. **Wong, N. P., Damico, J. N., and Salwin, H.,** Investigation of volatile compounds in codfish by gas chromatography and mass spectrometry, *J. Assoc. Off. Anal. Chem.*, 50, 8, 1967.
39. **Hrdlicka, J. and Kuca, J.,** The changes of carbonyl compounds in the heat-processing of meat. II. Turkey meat, *Poult. Sci.*, 44, 27, 1965.
40. **Mookherjee, B. D., Deck, R. E., and Chang, S. S.,** Relation between monocarbonyl compounds and flavor of potato chips, *J. Agric. Food Chem.*, 13, 131, 1965.
41. **Staeuble, E. J. and Rast, D.,** Readily volatile material in *Agaricus bisporus, Experientia,* 27, 886, 1971.
42. **Gianturco, V. M., Giammarino, A. S., and Friedel, P.,** Volatile constituents of coffee, *Nature,* 210, 1358, 1966.
43. **Grey, T. C. and Shrimpton, D. H.,** Volatile components of raw chicken breast muscle, *Br. Poult. Sci.*, 8, 23, 1967.
44. **Sato, Y., Watanabe, K., and Tanaka, Y.,** Chemical studies on odiferous compounds in hen egg. I. Volatile carbonyl and basic compounds in egg white, *Agric. Biol. Chem.*, 32, 405, 1968.
45. **Kahn, J. H., LaRoe, E. G., and Conner, H. A.,** Whiskey composition: identification of components by single-pass gas chromatography – mass spectrometry, *J. Food Sci.*, 33, 395, 1968.
46. **Bourne, W.,** On the effects of acetaldehyde, ether peroxide, ethyl mercaptan, ethyl sulphide and several ketones – di-methyl, ethyl methyl and di-ethyl – when added to anesthetic ether, *Anesth. Analg.*, 6, 131, 1927.
47. **Bourne, W.,** On the effects of acetaldehyde, ether peroxide, ethyl mercaptan, ethyl sulphide, and several ketones – di-methyl, ethyl methyl and di-ethyl – when added to anesthetic ether, *J. Pharmacol. Exp. Ther.*, 28, 409, 1926.
48. **Nakaaki, K.,** An experimental study on the effect of exposure to organic solvent vapor in human subjects, *J. Sci. Labour Part 2,* 50, 89, 1974 (Japanese).
49. **Wurster, D. E. and Munies, R.,** Factors influencing percutaneous absorption. II. Absorption of methyl ethyl ketone, *J. Pharm. Sci.*, 54, 554, 1965.
50. **Malten, K. E., Spruit, D., Boemaars, H. G. M., and de Keizer, M. J. M.,** Horny layer injury by solvents, *Berufs Dermatosen,* 16, 135, 1938.
51. **Tada, O., Nakaaki, K., and Fukabori, S.,** An experimental study on acetone and methyl ethyl ketone concentrations in urine and expired air after exposure to those vapors, *J. Sci. Labour Part 2,* 48, 305, 1972 (Japanese).
52. **Cushny, A. R.,** On the exhalation of drugs by the lungs, *J. Physiol.* (London), 40, 17, 1910.
53. **Winklhofer, F.,** Uber den Abbau aliphatischer Ketone durch uberlebendes Gewebe, *Z. Physiol. Chem.*, 263, 235, 1940.
54. **Leibman, K. C.,** Reduction of ketones in liver cytosol, *Xenobiotica,* 1, 97, 1971.
55. **Kimura, E. T., Ebert, D. M., and Dodge, P. W.,** Acute toxicity and limits of solvent residue for sixteen organic solvents, *Toxicol. Appl. Pharmacol.*, 19, 699, 1971.
56. **Divencenzo, A. and Krasavage, W. J.,** Serum ornithine carbamyl transferase activity as a liver response test for exposure to organic solvents, *Am. Ind. Hyg. Assoc. J.*, 35, 21, 1974.
57. **La Belle, C. W. and Brieger, H.,** The vapor toxicity of a composite solvent and its principal components, *Arch. Ind. Health,* 12, 623, 1955.
58. **Carpenter, C. P., Smyth, H. F., and Pozzani, U. C.,** The assay of acute vapor toxicity, and the grading and interpretation of results on 96 chemical compounds, *J. Ind. Hyg. Toxicol.*, 31, 343, 1949.
59. **Duckett, S., Williams, N., and Francis, S.,** Peripheral neuropathy associated with inhalation of methyl-*n*-butyl ketone, *Experientia,* 30, 1283, 1974.
60. **Verstraete, A., Van Der Stock, J., and Mattheeuws, D.,** Methylethylketonvergiftiging bij de hond, *Verh. K. Acad. Geneeskd. Belg.*, 26, 71, 1964.
61. **Schutz, E.,** Wirkungen organischer Flussigkeiten auf die Haut, *Arzneim. Forsch.*, 10, 1027, 1960.
62. **Golubev, A. A.,** Changes in the diameter of the rabbit pupil under the action of some industrial poisons producing irritation, *Gig. Tr. Prof. Zabol.*, 13, 58, 1969 (Russian).
63. **Larson, P. S., Finnegan, J. K., and Haag, H. B.,** Observations on the effect of chemical configuration on the edema-producing potency of acids, aldehydes, ketones, and alcohols, *J. Pharmacol. Exp. Ther.*, 116, 119, 1956.
64. **Schwetz, B. A., Leong, B. K. J., and Gehring, P. J.,** Embryo- and fetotoxicity of inhaled carbon tetrachloride, 1,1-dichloroethane and methyl ethyl ketone in rats, *Toxicol. Appl. Pharmacol.*, 28, 452, 1974.
65. **Holmberg, B. and Malmfors, T.,** The cytotoxicity of some organic solvents, *Environ. Res.*, 7, 183, 1974.
66. **Tionov, L. A., Smirnova, O. I., Kachurina, N. A., and Men'shikova, A.,** The radiosensitizing effect of methylethylketone diperoxide during x-ray irradiation of albino rats with lymphosarcoma, *Med. Radiol.*, 12, 91, 1967 (Russian).

67. **Lindenberg, A. -B. and Massin, M.,** Pouvoir cytolytique des alcools, des cetones et des phenols chez la levure de boulangerie, *J. Physiol.,* (Paris), 49, 285, 1957.
68. **Finney, D. J.,** *Probit Analysis,* 3rd ed., Cambridge University Press, 1971.
69. **Aviado, D. M.,** *Sympathomimetic Drugs,* Charles C Thomas, Springfield, Illinois, 1970, chap. 11.
70. **Aviado, D. M., Zakhari, S., Simaan, J. A., and Ulsamer, A. G.,** *Methyl Chloroform and Trichloroethylene in the Environment,* Part 1, Methyl Chloroform, CRC Press, 1976, p. 31–36.
71. **Simaan, J. A. and Aviado, D. M.,** Hemodynamic effects of aerosol propellants. I. Cardiac depression in the dog, *Toxicology,* 5, 127, 1975.

# *Part 3*
# *Methyl Isobutyl Ketone*

Chapter 10

# REVIEW OF THE LITERATURE ON METHYL ISOBUTYL KETONE

Methyl isobutyl ketone (MIK) is a congener of acetone, with many properties similar to methyl ethyl ketone (see Chapters 6 to 9). It has found extensive use in industry as a solvent, as well as in consumer products (aerosols, cleaning solutions, household cements).[1]

The literature concerning the physical, chemical and biological properties of MIK is reviewed here to serve as an introduction to the remaining chapters. A conversion table is presented at the end of the review to enable conversion of vapor concentrations in ppm, percent (v/v), mg/m$^3$, and mg/l. Topics in the review are discussed under the following headings: (A) Chemical and physical properties, (B) Medical uses, (C) Industrial uses, (D) Consumer products, (E) Human investigation, (F) Toxicologic investigation in animals, (G) Pharmacodynamics in man and animals, (H) Threshold limit values, and (I) Conversion table.

### A. Chemical and Physical Properties

Synonyms: 4-methyl-2-pentanone, 2-methyl-4-pentanone, methyl isobutyl ketone, isobutyl methyl ketone, isopropyl acetone, Hexone, MIBK; the symbol MIK is used in this monograph.

Structural formula:

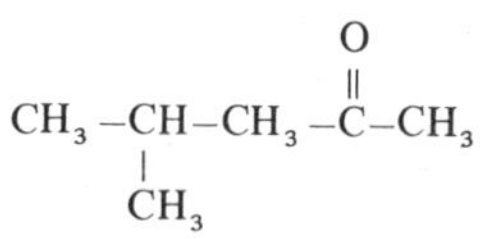

C = 71.94%
H = 12.08%
O = 15.97%

Molecular formula: $C_6H_{12}O$

Molecular weight: 100.09.

Boiling range (commercial product): 116 to 118°C; boiling point (pure MIK): 119°C.

Melting point: -84.7°C.

Specific gravity: 0.801 (4°C to 20°C); 0.798 (20°C).

Refractive index: 1.395 at 20°C.

Vapor pressure: Conflicting values, perhaps due to presence of water and other contaminants.
- 10 mm Hg at 30°C[2]
- 25 mm Hg at 20°C[3]
- 25 mm Hg at 30°C (manufacturer)
- 26.2 mm Hg at 30°C[4]
- 40 mm Hg at 29.5°C[5] (pure MIK).

Saturated vapor: 2.8% (v/v), 114.8 mg/l at 25°C (approximate value).

Solubility: 1.91% in water (slight to moderate solubility); soluble in all proportions in alcohol, benzene, ether, and acetone; also soluble in chloroform.

Colorless liquid, faint ketonic and camphor odor.

Flash point: 22.8°C.

Preparation: produced commercially by reduction (hydrogenation) of mesityl oxide (4-methyl-3-penten-2-one)

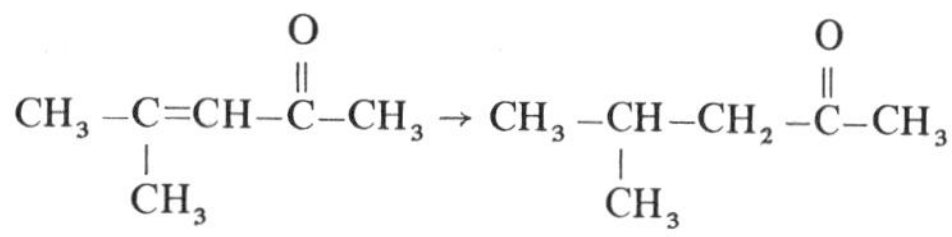

Determination of MIK: For measurement of MIK in air, Specht[5] uses a modification of the iodine titration method for ketones of Patty and co-workers.[6] Sodium hydroxide (1 *N*, 50 ml) is placed in a 1000-ml flask. The sample is introduced and the flask shaken to absorb the MIK. An excess of 0.1 *N* iodine solution is then introduced. After cooling the flask contents to room temperature, 25 ml of $H_2SO_4$ (2 *N*) are added. The excess iodine is then titrated with 0.1 N sodium thiosulfate, with starch solution used to determine the end point. This method gives an average recovery of 104%.

Kovacs[7] has employed a colorimetric technique for the same purpose. Air, 1 to 2 l, is bubbled through two impingers in series, each containing 4 ml $H_2O$. Solutions are transferred to a test tube, the volume brought to 10 ml, and 3% furfural and 1 ml 50% NaOH added. After 1 hr, the tube is cooled and 5 ml $H_2SO_4$ added drop by drop. After mixing, the absorbancy is read at 530 nm.

Gas chromatography and mass spectroscopy have recently become popular for ketone (MIK) determination in biological fluids such as urine.[8]

### B. Medical Uses

MIK is employed in public health environmental studies for determining the presence of heavy metals in air as well as in biological materials. For example, lead in air[9] and in biological materials[10] can be extracted in MIK and analyzed by atomic absorption spectrophotometry. Such a method has been used in a mass screening program for the presence of lead in the children of Chicago.[11]

Similar techniques can be used for detecting antimony,[12] thallium,[13] and cadmium.[14]

### C. Industrial Uses

MIK is used for similar purposes as methyl ethyl ketone, reported on previously from this laboratory (see Chapter 6). Unlike acetone, which has also found extensive use as a chemical intermediate, such as in the production of methyl methacrylate,[15] the popularity of these larger ketones is primarily for their solvent properties. (Comparative physical properties of acetone, methyl ethyl ketone, and MIK have been presented in graphic form by Gallant[15].) Originally used, for the most part, in making lacquers and varnishes,[5] industrial and home applications of non-acetone ketones and their products have steadily grown. By 1968, yearly consumption of MIK was 200,000,000 lb and methyl ethyl ketone, 300,000,000 lb.[15]

In a 1953 monograph, Browning[4] reported the chief industrial application of MIK as an agent to reduce the viscosity of cellulose lacquers. In general, MIK was found useful as a solvent for nitrocellulose, and as a replacement for acetone, which is not completely able to dissolve certain substances such as ester gum and some phenol-type resins.

In a comparison of the solvent powers and dilution ratios of acetone and other ketones, Sevestre[16] found two basic advantages of MIK and methyl ethyl ketone over substances with the same evaporation rate: (a) lower viscosity solutions for the same solids concentration, and (b) very high hydrocarbon dilution ratio.

More recently, plastics manufacturing has employed MIK in a variety of ways. MIK and methyl ethyl ketone are recommended by Sevestre[17] for use in polyurethane coating, based on criteria of solution viscosity and water absorption. MIK is also used as a solvent in the production of vinyl and acrylic resins.[18] Among solvents used for forming monomolecular films, MIK was found superior to methyl ethyl ketone in solvent power, although the latter withstood a higher force at break.[19] Application has also been found for MIK in repairing surface defects in thermoplastics.[20]

In addition to the uses described above, MIK has been proposed as a diluting and denaturing agent for ethyl alcohol.[3]

Finally, the petroleum industry utilizes MIK in the dewaxing of oils[21] as well as in removing metals from lubricating oils.[22]

### D. Consumer Products

Gleason et al.[23] report the presence of MIK in several consumer products. The composition for some but not all of the products is known. In the following list, the percentage composition is expressed as v/v

1. Spray paint (one product, concentration unknown), tractor and implement finish (three products), and marine engine finish (one product).
2. Products containing SDA No. 1 solvent formula: four products for fuel, metal cleaning, shellac, and thinner, as well as one for rotogravure printing operations. All contain SDA No. 1 solvent formula (composition: 20 parts ethyl alcohol, 0.8 parts methanol, and 0.2 parts MIK). Additional MIK (approximately 1%) is also contained in these products.
3. Products containing industrial solvent No. 4: primer-sealer, black gloss paint, and shellac (one product each) contain in various amounts up to 70% industrial solvent No. 4. This is composed of ethyl alcohol 80%, methyl alcohol 4%, isopropyl alcohol 15%, and MIK 1%.
4. Other products: airplane cement and adhesive (one product, MIK concentration unknown); gasket sealer, contains ether isopropyl alcohol or MIK (one product, concentration unknown); proprietary solvent (one product contains 21% MIK); industrial solvents (12 products, 1% MIK); professional insecticide (one product, 42.5% MIK); insecticide for ornaments and trees (one product, 40% MIK); liquid solder (one product, concentration unknown); lacquer thinner (one product, concentration unknown); one Swedish product, 10% MIK)[24]; anti-fouling paint for ships (one product, concentration unknown)[25]; low-toxicity fishnet coating[26] (one product, concentration unknown).

### E. Human Investigation

There is no information on the potential toxicity of MIK contained in consumer products. Most of the available literature is related to occupational exposure to MIK.

#### *1. Occupational Hazards*

MIK is not considered particularly hazardous[27]

(no deaths reported), although it is regarded by Gleason et al.[23] as probably more toxic than methyl ethyl ketone. Elkins[27] reports two separate incidents of complaints by workers exposed to the current TLV[28] of 100 ppm (0.41 mg/l). In the first, while waterproofing boots, employees noted development of headache and nausea. For the second group, the only symptom was respiratory tract irritation. In both groups, effects were reduced as the work week progressed, suggesting increased tolerance to MIK. Further evidence of tolerance was the reoccurrence of symptoms at the beginning of the following week. Most MIK effects were abolished when the ambient concentration was reduced to 20 ppm.

In a 1956 compendium of toxic effects of industrial solvents, Smyth[29] notes that "simple ketones are irritant and narcotic but not chronically toxic." This conclusion may be open to question, in light of a 1968 study in a Milanese factory. Armeli and co-workers[30] documented symptoms of workers exposed over a 5-year period, initially to a concentration ranging from 80 to 500 ppm (0.33 to 2.1 mg/l) for 30 min per day, and later from 50 to 100 ppm (0.21 to 0.41 mg/l). The most common symptoms after this period in the 14 workers involved were asthenia, insomnia, headache, or sleepiness (total six workers), and diarrhea, vomiting, and other intestinal disorders (total five workers). By contrast, irritation of the conjunctival or nasal mucosae, common effects of MIK, occurred in only three workers. These findings are similar to the report of Elkins[27] in that major toxic effects centered on the central nervous system and gastrointestinal system. Unlike the experience of Elkins[27] and conclusion of Smyth,[29] however, the 5-year duration of effects would appear to belie the concept of immunity or tolerance to MIK. Since these reports (Elkins[27] and Smyth[29]) form part of the basis of the current TLV, re-examination of long-term effects of MIK would seem to be warranted.

Other problems caused by exposure to MIK include irritation of the eye[5,31] and nasal mucosa,[4] as well as an objectionable odor.[28,32] Although the TLV recognizes difficulties with the odor at concentrations of 100 ppm (0.41 mg/l) or greater, May[32] found that human subjects noticed the odor at concentrations as low as 8 ppm (0.033 mg/l).

*2. Addiction and Abuse*

Abuse of MIK may occur through inhalation of solvents by "sniffers." The chemical composition of at least one thinner product, for example, includes:

acetates (ethyl, propyl, butyl) 15 to 25%;
aliphatic alcohols (ethyl, propyl, butyl) 10 to 50%;
toluene 35 to 70%;
MIK 10%.

Nylander finds that the clinical signs of acute thinner intoxication resemble acute alcohol intoxication.[24] However, it is also possible that certain symptoms such as somnolence, and more severely, unconsciousness, might in part be due to MIK.

*3. Experimental Inhalation*

Based on human sensory response studies by Silverman and co-workers,[33] and single-dose inhalation experiments in animals, Smyth[29] predicted the following toxic effects of MIK on humans exposed to the TLV and multiples of the TLV for an 8-hr workday. While narcosis is the most important effect of MIK inhalation, the odor of MIK was predicted the most likely problem for anyone exposed for 8 hr to the TLV (100 ppm or 0.41 mg/l). Exposure to twice the TLV (200 ppm or 0.82 mg/l) was proposed to cause irritation of the throat as well as of conjunctival and nasal mucous membranes. Central nervous system depression, i.e., narcosis, was an additional toxic effect occurring at ten times the TLV (1000 ppm or 4.1 mg/l). In general, these toxicities have been observed by workers exposed to MIK. However, industrial experience also indicates that gastrointestinal disturbances, manifested as nausea, vomiting, loss of appetite, and other disorders,[27,30] might have been included, perhaps at one to five times the TLV.

Some of the estimations for graded effects of MIK cited by Smyth[29] no doubt were derived from the sensory response experiments of Silverman and associates.[33] These investigators exposed 12 subjects to solvent vapors for a period of 15 min. A motion picture was shown during the exposure period to distract the subjects from focusing on the vapor effects. A majority of subjects were found to respond to the odor of MIK at 100 ppm (0.41 mg/l). Irritating effects to the eyes, nose, and throat were felt with concentrations greater than 200 ppm (0.82 mg/l). The conclusion of most subjects, that 100 ppm (0.41 mg/l) was the highest satisfactory concentration for an 8-hr workday, has since become part of the basis for the current TLV of 100 ppm (0.41 mg/l).[28]

Somewhat different experiments than those of Silverman and co-workers[33] have been carried out in more recent times. In a study by May[32] of odor threshold, varying vapor concentrations were generated in test flasks and sniffed (random order). Gas chromatography of vapor or aliquots allowed accurate determination of concentrations. Results show an odor threshold (first perceptible concentration) of 8 ppm (0.033 mg/l), while 15 ppm (0.062 mg/l) was clearly noticeable. These very low concentrations indicate, in contrast to Silverman et al.,[33] that response to the odor of MIK can occur at far below the TLV of 100 ppm.[28]

The potential eye-irritating effects of various solvents have also been investigated under carefully controlled conditions by Wayne and Orcutt.[31] Two test concentrations of various solvent vapors, 5 ppm (0.021 mg/l) and 20 ppm (0.082 mg/l), were directed to an eye mask worn by the subject (eight to eleven subjects). Subjects found that MIK had the greatest eye-irritating property of the solvents studied. Lesser irritation was caused by (in order) trichloroethylene, xylene, methyl ethyl ketone, hydrocarbon fractions, ethanol, isopropanol, and acetone. The authors concluded that eye-irritating potential is dependent on the composition of the solvent rather than its boiling point. From these studies by May[32] and Wayne and Orcutt,[31] it seems apparent that measurement of threshold effects using sophisticated techniques may ultimately reveal a far greater sensitivity to solvent vapors, including MIK, than previously concluded.

*4. Urinary Excretion*

In studies using 24-hr urine samples from men and women, Zlatkis and Liebich[8] found that ketones are among the most characteristic constituents of the volatile component of urine. Analysis by gas chromatography and mass spectroscopy revealed the presence of MIK, as well as other substituted 2-pentanones.

**F. Toxicologic Investigation**

In this section, toxicities to MIK are described on the basis of species, route of administration, and single or repeated exposures.

*1. Oral Toxicity: Single Administration in Rats and Mice*

On the basis of a range-finding test, Smyth and associates[34] noted an $LD_{50}$ of 2.08 g/kg (95% confidence range: 1.91 to 2.27 g/kg) in rats 14 days after a single, oral dose of MIK. Batyrova[35] found an oral $LD_{50}$ of 2.85 g/kg for mice and 4.6 g/kg for rats.

*2. Oral Toxicity: Repeated Administration in Rats*

Quinby[36] has compiled evidence showing that exposure to a variety of ketones may cause polyneuritis. He cites a recent work[36a] in which commercial-grade MIK in drinking water (520 mg/kg/ day for 5 months) induced "large numbers of dilated, glycogen-filled mitochondrial remnants, adaxonal Schwann cell invaginations, and rare focal swellings." This was characterized as "silent" neuropathy, in the absence of frank distal nerve fiber degeneration. These effects may have been due to the presence of 3% methyl *n*-butyl ketone, a known neurotoxic solvent, as a contaminant in the commercial-grade MIK.

*3. Inhalation Toxicity: Single Exposure in Rats and Mice*

Smyth and co-workers[34] exposed rats to saturated MIK vapor until at least one death occurred, as well as to 2000 and 4000 ppm (8.2 and 16.4 mg/l) for 4 hr. Results show that saturated vapor killed at least one rat within 15 min. No deaths occurred in six rats following 4 hr of exposure to 2000 ppm (8.2 mg/l), while all animals (six out of six) succumbed from exposure to 4000 ppm (16.4 mg/l) for 4 hr. Investigators for Shell Chemical Corporation found mice able to withstand 30-min exposures to 19,500 ppm (80 mg/l).[37]

*4. Inhalation Toxicity: Single Exposure in Guinea Pigs*

The most detailed account of acute toxicity to MIK vapor is reported by Specht[5] and Specht and co-workers[38] for guinea pigs. Symptoms were recorded during exposure to concentrations ranging from 0.1% (4.1 mg/l) to 2.8% (saturated, 114.8 mg/l) MIK, ten guinea pigs per concentration. Post-mortem examination was carried out to reveal additional damage to individual organs.

Symptoms (in order of increasing sensitivity in each category) were

1. Blinking of eyes, rubbing of nose, and lacrimation (irritation of the cornea and conjunctiva).

2. Sneezing, coughing, salivation, retching, rubbing of nose and mouth (irritation of buccal, nasal, and pharyngeal passages).

3. Postural instability, loss of corneal and auditory reflexes, change in respiratory rate and character, coma, and death (narcosis, central nervous system toxicity).

4. Decreased rectal body temperature, pulse rate (change in metabolism of tissues).

Data summarized for the lowest and highest concentrations are reported below, as well as a synopsis of events with an intermediate concentration, 1.68% (68.9 mg/l). Exposure to 0.10% (4.1 mg/l) caused little or no irritation of the eyes and nose of the guinea pigs (the operator *was* affected, however). Low-grade narcosis was indicated by a slight effect on maintenance of reflexes and stance, and a fall in respiratory rate (from about 140 to 70/min) during the first 6 hr of exposure. Body (rectal) temperature was also slightly reduced from 39°C to 37°C. There were no deaths overnight.

Exposure to saturated vapor, 2.8% (114.8 mg/l), caused the most severe symptoms, although of a shortened duration due to death of half the animals in 45 min and most in 60 min. The decline in respiratory rate and body temperature was at a much faster rate. Gasping-type breathing was also noted at this concentration.

Events occurring during exposure to intermediate dose, 1.68% (68.9 mg/l), were described in detail by the author. They are reported here in their entirety:

Up to 1 min – Immediately upon introduction to the test atmosphere the pigs began to squint and rub their noses and eyes violently. Copious lacrimation occurred and sneezing and coughing was very frequent.

2 min – Salivation was evident by the wet area about the mouth, neck, and brisket.

10 min – All pigs were so weak that they could not remain erect; other symptoms in proportion.

15 min – All pigs over on side making only feeble movements at most and usually quite inert. Corneal reflex still present. Pinnal flick in response to sharp click still present.

25 min – Respiratory rate has dropped to 35/min, rectal temperature to 36.6°C.

40 min – Pulse by palpation estimated at 240 beats/min. Auditory and corneal reflexes gone. Respiration gasping and random pawing movements appear occasionally.

60 min – Rectal temperature 34.7°C. Pulse 181 beats/min. Concentration 1.70%.

85 min – First death. Average temperature of survivors 33.4°C, pulse 143 beats/min, respiratory rate 20 breaths/min.

94 min – Second death.

108 min – Third death.

125 min – Fourth and fifth deaths. Average rectal temperature of survivors 32°C, pulse 155 beats/min, respiratory rate 85/min.

131 min – Sixth and seventh deaths.

134 min – Eighth death.

142 min – Ninth death. Concentration 1.53%.

204 min – Chamber opened, one pig still gasping 100 breaths/min, pulse 158 beats/min, rectal temperature 30.4°C. Death occurred at least 2 hr after the end of the exposure.

It might be noted that the fall in body temperature was of a similar rate (about 6°C in 2 hr) as that of a fresh guinea pig killed at the beginning of the experiment (about 8°C in 2 hr). This suggests a severe reduction in tissue metabolism caused by MIK.

Post-mortem examination revealed only slight changes in pigs succumbing to MIK, mostly confined to some indication of congestion in the brain and lungs. Scattered fatty metamorphosis could be seen microscopically in the liver. While there was some hypertrophy and congestion of the spleen, no damage was observed to the heart or kidneys.

### *5. Inhalation Toxicity: Repeated Exposure in Rats*

Vernot et al.[39] and MacEwen and associates[40] report that rats continuously exposed to MIK vapor, 410 mg/m$^3$ (0.41 mg/l), had significantly increased kidney weights and kidney/body weight ratios after 2 weeks, as well as kidney damage, described by MacKenzie[41] as hyaline droplet toxic nephrosis. These changes persisted if exposure was extended to 90 days, but even after this duration could be reversed by removal of the ketone vapors. Liver weight/body weight ratios also increased after 90 days exposure to 410 mg/m$^3$ (100 ppm), or 2 weeks exposure to 820 mg/m$^3$ (200 ppm). It is interesting that no other species similarly tested (monkeys, dogs, or mice) developed organ toxicity to MIK.

Exposure of rats to concentrations ranging from 86 to 127 mg/m$^3$ (21 to 31 ppm) for 4 hr per day, for 4.5 months, was found by Batyrova[35] to disturb conditioned reflexes and liver detoxification function, as well as to elevate the eosinophil count in the blood. On this basis, a maximum permissible concentration of MIK was suggested as 5 mg/m$^3$ (1.2 ppm).

*6. Inhalation Toxicity: Repeated Exposures in Monkeys, Dogs, and Mice*

Inhalation experiments by Vernot and coworkers[39] and MacEwen and associates[40] demonstrated damage in rats after 2 weeks exposure to 410 mg/m$^3$ (100 ppm), but MIK had no observable effect on mice, monkeys, and dogs.

**G. Pharmacodynamics**

In this section, toxicities caused by MIK are described by organ systems affected.

*1. Central Nervous System Toxicity*

Narcosis is believed the most important manifestation of toxicity due to MIK.[29] In a study following a 5-year industrial exposure ranging from 50 to 500 ppm (0.21 to 2.1 mg/l) for 30 min per day, the most common complaints reflected central nervous system toxicity (anorexia, asthenia, headache, insomnia, and sleepiness).[30] Elkins[27] reports headache as a primary symptom in one set of workers exposed to 100 ppm (0.41 mg/l), but not in another group similarly exposed.

In experiments using rats, Batyrova[35] found disturbances to conditioned reflexes following exposure to MIK, 86 to 127 mg/m$^3$ (21 to 31 ppm) for 4 hrs per day, 4.5 months. Specht[5] observed narcosis to a slight degree following inhalation of 0.1% (4.1 mg/l), and in increasing severity with graded concentrations (symptoms: postural instability, loss of corneal reflex and auditory reflex, change in respiratory rate and character, coma, death).

*2. Peripheral Nervous System Toxicity*

In a personal communication, Quinby[36] discussed the possibility that prolonged exposure to a variety of ketones, in paints for example, may cause polyneuritis in man. This report notes that chronic, oral administration in rats caused "silent" neuropathy.[36a] This pathologic effect in the most distal portions of the rats' extremity nerves was described as "large numbers of dilated, glycogen-filled mitochondrial remnants, adaxonal Schwann cell invaginations, and rare focal swellings . . . (without) frank distal nerve fiber degeneration . . . ." Neuropathy may have been due to 3% methyl *n*-butyl ketone, found as a contaminant in the commercial-grade MIK.

*3. Cardiovascular Toxicity*

Specht[5] and Specht and associates[38] noted a decline in pulse rate from 300 to 85 beats/min in guinea pigs exposed to 1.68% (68.9 mg/l) MIK. Toward the end of the experiment, irregularities in the heart beat were also observed. On post-mortem examination, there was no evidence of cardiac abnormalities in pigs succumbing to MIK.

The authors are unaware of any studies concerning hemodynamic effects of MIK administered by any route.

*4. Pneumotoxicity*

Guinea pigs exhibited a graded decrease in respiratory rate during inhalation of 0.1% to 2.8% (4.1 to 114.8 mg/l) MIK. A fall from 117 breaths/min (control) to 35 breaths/min was observed during exposure to an intermediate concentration, 1.68% (68.9 mg/l). At higher concentrations, the slow respiratory rate was replaced by gasping-type breathing. This response was believed due to central nervous system toxicity. Post-mortem examination revealed slight lung congestion in animals which died.[5,38]

*5. Oculotoxicity*

Irritation of the eyes is a common effect of exposure to MIK. Silverman and associates[33] report that a majority of subjects complained of eye irritation during 15-min exposure to 200 ppm (0.8 mg/l). Using somewhat more controlled conditions, Wayne and Orcutt[31] found early indication of eye irritation during exposure to 5 ppm (0.021 mg/l) and 20 ppm (0.082 mg/l). Industrial exposure of 13 workers to concentrations ranging from 50 ppm (0.21 mg/l) to 500 ppm (2.1 mg/l) over a 5-year period (30 min per day), produced only one complaint of eye irritation.[30]

In animal experiments, Specht[5] noted corneal and conjunctival irritation as an early response of guinea pigs to MIK inhalation. This was exhibited as blinking of eyes, lacrimation, and rubbing of eyes. It is noteworthy that although animals were not affected by the lowest concentration, 0.1% (4.1 mg/l), the operator became extremely uncomfortable.

*6. Nasopharyngeal Toxicity*

The odor of MIK is the most noticeable

characteristic of the solvent at its TLV of 100 ppm (0.41 mg/l), according to Smyth.[29] This is based on human sensory response experiments by Silverman and co-workers,[33] in which a majority of subjects found the odor objectionable at 100 ppm (0.41 mg/l). A later study by May[32] revealed a very low odor threshold, MIK becoming recognizable at only 8 ppm (0.033 mg/l) and clearly defined at 15 ppm (0.062 mg/l).

Irritation of the mucous membranes of the eyes, nose, and throat is considered by Smyth[29] the major additional effect (besides odor) of exposure to 200 ppm (0.82 mg/l). Elkins,[27] however, reports that workers in one plant complained of respiratory tract irritation during exposure to only 100 ppm (0.41 mg/l).

Guinea pigs subjected to MIK vapor exhibit signs of irritation of the mouth, nose, and throat at higher concentrations or longer exposure times than irritation of the eyes. Symptoms of these effects are sneezing, coughing, salivation, retching, and rubbing of the nose and mouth.[5,38]

*7. Renal Toxicity*

In studies of rats, dogs, monkeys, and mice, Vernot et al.[39] and MacEwen and co-workers[40] found that rats were the only species to develop kidney damage. Two weeks of exposure to 410 mg/m$^3$ (100 ppm) induced increases in kidney weight and kidney weight/body weight ratios. White dots appearing on the kidney surface were revealed by microscopic examination to resemble hyaline droplet tubular nephrosis. MacKenzie[41] interpreted these dots to represent a swelling of the proximal tubules, believed caused either by an increase in protein leakage through the glomerular membrane or an inability of the tubular cells to remove the protein absorbed. These effects were able to be reversed, even after 90 days exposure, by removal of the animal from the vapor. In contrast to these findings in rats, Specht[5,38] found no kidney changes in guinea pigs succumbing to MIK exposure, 0.1% to 2.8% (4.1 to 114.8 mg/l).

*8. Hepatotoxicity*

Vernot et al.[39] and MacEwen and associates[40] reported an increase in liver weight and liver weight/body weight ratios in rats exposed to 820 mg/m$^3$ (200 ppm), for 2 weeks of 410 mg/m$^3$ (100 ppm) for 90 days. Batyrova[35] found that rats exposed to 86 to 127 mg/m$^3$ (21 to 31 ppm) developed a deficiency in the detoxifying function of the liver. Fatty metamorphosis was observed in scattered hepatocytes from guinea pigs succumbing to MIK.[5]

*9. Other Toxic Effects*

Armeli and associates[30] report two cases of questionable splenomegaly among 14 workers exposed to MIK vapors (30 min per day, 5 years) at concentrations of 50 to 500 ppm (0.21 to 2.1 mg/l). High eosinophil counts were found by Batyrova[35] in rats exposed to 86 to 127 mg/m$^3$ (21 to 31 ppm) for 4 hr per day, 4.5 months. Signs of metabolic depression have been observed by Specht[5] and Specht and co- workers[38] in guinea pigs exposed to 0.1% to 2.8% (4.1 to 114.8 mg/l) MIK. Decreased metabolic function was revealed by a decline in pulse rate and body temperature, the latter approximating the fall in temperature of a freshly killed guinea pig.

## H. Threshold Limit Value

The TLV[28] of 100 ppm (0.41 mg/l) is based partly on the human sensory response test of Silverman and others,[33] in which subjects estimated 100 ppm (0.41 mg/l) to be the maximum tolerable concentration for an 8-hr period. Elkins[27] report is also noted, in which workers complained of various symptoms (headache, nausea, and respiratory tract irritation) after exposure to 100 ppm. However, the development of tolerance cited by Elkins appears to have been a mitigating factor. Experimental animal studies used in establishing the TLV include the findings of Specht[5] in guinea pigs exposed to 0.1% to 2.8% (4.1 to 114.8 mg/l) as well as the report of Smyth and others,[34] that rats succumbed to 4000 ppm (16.4 mg/l) but not to 2000 ppm (8.2 mg/l). Prior to this TLV, Cook[42] believed 200 ppm (0.82 mg/l) safe over an 8-hr period, based on an arbitrary fraction of the threshold concentration (0.1%, 4.1 mg/l) for effects in guinea pigs as reported by Specht.[5]

## I. Conversion Table

Since vapor concentration of MIK can be expressed by several methods, a table to enable conversion of units of concentrations is presented on the next page.

| v/v% | ppm | mg/m$^3$ | mg/l |
|---|---|---|---|
| 0.0001 | 1 | 4.1 | 0.0041 |
| 0.00024 | 2.4 | 10.0 | 0.01 |
| 0.001 | 10 | 41.0 | 0.041 |
| 0.0024 | 24 | 100 | 0.1 |
| 0.01 | 100 | 410 | 0.41 |
| 0.0244 | 244 | 1000 | 1.0 |
| 0.1 | 1000 | 4100 | 4.1 |
| 0.2443 | 2443 | 10,000 | 10.0 |
| 1.0 | 10,000 | 41,000 | 41.0 |
| 2.8 (saturated) | 28,000 | 114,800 | 114.8 |

Chapter 11

# ACUTE ORAL, INTRAPERITONEAL, AND INHALATIONAL TOXICITY OF METHYL ISOBUTYL KETONE IN THE MOUSE

Acute toxicity of MIK has been studied in various species after oral[34-36] and inhalational[5,34,35,37-41] but not after parenteral administration (see Chapter 10). Furthermore, no investigation has been undertaken to elucidate the toxicity of MIK in the same animal species after administration by various routes. The primary objective of this study is, therefore, to determine the acute toxicity of MIK in mice following administration via oral, intraperitoneal, and inhalational routes. A comparison of the toxicity of methyl isobutyl ketone, methyl ethyl ketone, and acetone is also reported. The correlation between toxicity and molar refractivity is uncovered and reported for the first time.

## A. Methods

Experiments were performed on male mice (CF-1 strain, Charles River Laboratories, Wilmington, Massachusetts). Certified MIK from Fisher Scientific Company has been used. Mice weighing from 20 to 23 g were used for inhalational and intraperitoneal adminstration, and those weighing from 28 to 34 g were given MIK orally. The mean weight of mice was approximately the same in all groups receiving MIK via the same route of administration. Mice were divided into three series as follows: oral administration of 0.9, 1.5, 2.0, 3.0, and 3.5 g/kg; intraperitoneal injection 0.25, 0.38, 0.50, 0.75, 1.00, and 1.25 g/kg; and inhalation of various concentrations for 45 min. In this last series, groups of mice of ten animals each were introduced into a 10-liter glass chamber and various concentrations of MIK in air were flushed into the chamber at a rate of 2 to 3 l/min. Mortality rate was calculated after a 45-min exposure period. This exposure time was chosen to allow for comparison of the toxicity of MIK with that of methyl ethyl ketone reported previously (Chapter 7).

Comparison of the toxicity of the studied ketones revealed that MIK is more toxic than methyl ethyl ketone. To find out any correlation between the physical properties of the molecule and toxicity, it was found necessary to determine the toxicity of the lower homologue, viz., acetone. The intraperitoneal $LD_{50}$ of acetone was, therefore, determined using the following dosage levels: 1.576, 2.364, 3.152, 3.940, 4.728, and 5.516 g/kg.

The mortality rate was calculated after 24 hr in the case of mice that received MIK orally and intraperitoneally, and 45 min after the inhalation of MIK vapors. $LD_{50}$ and $LC_{50}$ (95% fiducial limits) were then determined according to the probit method.[43]

## B. Results

Data of acute toxicity experiments where MIK was given orally are summarized in Table 11.1. The dose that brought about 50% mortality was $1.90 \pm 0.68$ g/kg.

Table 11.2 shows that the intraperitoneal dose of MIK associated with 50% death after 24 hr is $0.59 \pm 0.23$ g/kg.

Table 11.3 shows the results of inhalation toxicity after 45 min exposure of mice to MIK vapors. A saturated concentration of 3% MIK failed to bring about 100% mortality. It was associated with 80% deaths. On the other hand, a concentration of 1% MIK caused no mortality. The $LC_{50}$ (95% fiducial limit) is $74.2 \pm 25.8$ mg/l.

Comparison of the toxicity of MIK and methyl ethyl ketone following various routes of administration is shown in Table 11.4. Generally, MIK proved to be more toxic than methyl ethyl ketone by all routes of administration. Thus, the oral and intraperitoneal $LD_{50}$ calculated in terms of mMol/kg is 19.0 and 5.9 in the case of MIK, and 43.5 and 23.0 in the case of methyl ethyl ketone, respectively. Inhalational $LC_{50}$ is 2.84 and 0.74 mMol/l for methyl ethyl ketone and MIK, respectively. The ratio between the $LD_{50}$ of MIK and methyl ethyl ketone given intraperitoneally or by inhalation is 25.7% and 26.1%, respectively, while following oral administration, the ratio is 43.7%.

Table 11.5 summarizes the toxicity of acetone given by the intraperitoneal route; the $LD_{50}$ is $3.17 \pm 0.42$ g/kg.

## C. Discussion

The oral toxicity of MIK in rats was studied by Smyth et al.,[34] who reported an $LD_{50}$ of 2.08 g/kg 14 days after a single dose of MIK. On the other hand, Batyrova[35] found that 2.85 g/kg and 4.6 g/kg are lethal for 50% of mice and rats, respectively. In the present study, an oral $LD_{50}$ of

TABLE 11.1

**Acute Toxicity of MIK Given Orally to Mice**

| Group* | Dose (g/kg) | % Mortality | $LD_{50}$ (g/kg) 24 hr | Regression coefficient |
|---|---|---|---|---|
| 1 | 0.9 | 13 | | |
| 2 | 1.5 | 25 | | |
| 3 | 2.0 | 38 | | |
| 4 | 2.5 | 75 | 1.90 ± 0.68 | 0.9808 |
| 5 | 3.0 | 88 | | |
| 6 | 3.5 | 100 | | |

*Each group consisted of eight animals.

TABLE 11.2

**Acute Intraperitoneal Toxicity of MIK in Mice.**

| Group* | Dose (g/kg) | % Mortality | $LD_{50}$ (g/kg) 24 hr | Regression coefficient |
|---|---|---|---|---|
| 1 | 0.25 | 0 | | |
| 2 | 0.38 | 13 | | |
| 3 | 0.50 | 40 | 0.59 ± 0.23 | 0.9354 |
| 4 | 0.75 | 80 | | |
| 5 | 1.00 | 80 | | |
| 6 | 1.25 | 100 | | |

*Each group consisted of ten animals, except Groups 2 and 6 which each contained eight animals.

TABLE 11.3

**Acute Inhalational Toxicity of MIK in Mice after 45 min of Exposure**

| Group* | Concentration | | % Mortality | $LC_{50}$ | Regression coefficient |
|---|---|---|---|---|---|
| | % v/v | mg/l | | | |
| 1 | 1.00 | 41.0 | 0 | | |
| 2 | 1.25 | 51.3 | 20 | | |
| 3 | 1.50 | 61.5 | 50 | 74.2 ± 25.8 | 0.9506 |
| 4 | 2.00 | 82.0 | 60 | | |
| 5 | 3.00** | 123.0 | 80 | | |

*Each group consisted of ten mice.
**Saturated

1.90 g/kg (95% fiducial limits are 1.22 to 2.58 g/kg) was found. These differences are difficult to evaluate because of variation in species and duration of experiments in the case of the former authors[34] and because of the grade of MIK in the case of the latter.[35]

No intraperitoneal $LD_{50}$ has been reported in any species. Thus, the present investigation constitutes the first reported $LD_{50}$ for mice after intraperitoneal administration, 0.59 ± 0.23 g/kg.

Inhalational toxicity after a single exposure to MIK has been reported in rats,[34] mice,[34] and guinea pigs.[5,38] In mice exposure for 30 min to a concentration of 1.95% MIK (79.95 mg/l) was reported to induce no mortality. By comparison to these results, in the present study, the concentration of MIK that killed 50% of mice in 45 min was 1.81% (74.2 mg/l or 18,100 ppm). Smyth et al.[34]

found that 2.8% (28,000 ppm) killed at least one rat within 15 min.

Comparison of MIK and methyl ethyl ketone revealed that MIK is more toxic than the other ketone. In the case of intraperitoneal and inhalational routes of administration, the ratios between the toxic doses of methyl ethyl ketone and MIK are 25.7% and 26.1%, respectively. Orally administered ketones showed, however, a different ratio, 43.7%. This difference may reflect varying rates of intestinal absorption and/or rates of metabolism or excretion of these ketones.

Results of these acute intraperitoneal toxicity experiments of MIK and acetone, as well as those of methyl ethyl ketone (Chapter 7), demonstrate the following order of increasing toxicity among ketones: acetone < methyl ethyl ketone < MIK. In an attempt to evaluate this order of toxicity, it was considered appropriate to compare acute intraperitoneal toxicity with differences in physical properties of these substances such as molecular weight, density, and refractive index. These three properties may be combined to give a single calculated value as shown by the Lorentz-Lorenz equation,[49] which states that:

$$P_r = \frac{n^2 - 1}{n^2 + 2} \cdot \frac{m}{d}$$

where $P_r$ is the molar refractivity, n is the refractive index, m is the molecular weight, and d is the density. The molar refractivity of acetone is 16.23, that of methyl ethyl ketone is 20.81, and that of MIK is 30.05. The chemical structures of these ketones are as follows:

| | |
|---|---|
| Acetone: | $CH_3COCH_2-H$ |
| Methyl ethyl ketone: | $CH_3CO\overset{1}{C}H_2-\overset{2}{C}H_3$ |
| MIK: | $CH_3CO\overset{1}{C}H_2-\underset{CH_3}{\overset{2}{C}H}-CH_3$ |

The difference between the chemical structure of MEK and MIK is the substitution on the carbon atom adjacent to the carbonyl group ($c_1$); thus it is methyl in the case of methyl ethyl ketone and isopropyl in the case of MIK. The polarizability of the bond between $C_1$ and $C_2$ might, therefore, account for the difference in reactivity between methyl ethyl ketone and MIK. The polarizability of the $C_1-C_2$ bond in methyl ethyl ketone is 4.58, while that of the same bond in MIK is 13.82. The ratio between these two values is 33.19%, which seems to correlate with the ratio between toxicities of methyl ethyl ketone and MIK, viz., 25.7% (intraperitoneally) and 26.1% (by inhalation).

The correlation between molecular weight, refractive index, and density (as reflected from the molar refractivity) on one hand and the toxicity on the other is shown in Figure 11.1. The toxicity of methyl ethyl ketone and MIK was compared with that of acetone. A positive regression line was obtained which indicates that the polarizability of the molecule is a determinant factor in toxicity with these compounds. To find out whether or not

TABLE 11.4

**Comparative $LD_{50}$ and $LC_{50}$ of MEK and MIK in Mice Following Oral, Intraperitoneal, and Inhalational Routes of Administration**

| | $LD_{50}$ mMol/kg (24 hr) | | $LC_{50}$ mMol/l (45 min) |
|---|---|---|---|
| Ketone | Oral | Intraperitoneally | inhalation |
| MEK | 43.5 | 23.0 | 2.84 |
| MIK | 19.0 | 5.9 | 0.74 |
| MIK%/MEK | 43.7 | 25.7 | 26.1 |

TABLE 11.5

**Acute Intraperitoneal Toxicity of Acetone Administered to Mice**

| Group | No. of mice | Dose (g/kg) | % Mortality | $LD_{50}$ (g/kg) 24 hr | Regression coefficient |
|---|---|---|---|---|---|
| 1 | 12 | 1.576 | 0 | | |
| 2 | 10 | 2.364 | 20 | | |
| 3 | 10 | 3.152 | 50 | 3.17 ± 0.42 | 0.9909 |
| 4 | 10 | 3.940 | 70 | | |
| 5 | 9 | 4.728 | 88.9 | | |
| 6 | 9 | 5.516 | 100 | | |

this correlation occurs in other compounds, the relative toxicity of alcohols reported by Browning[4] is plotted against the calculated molar refractivity according to the equation mentioned above. A positive correlation was found between these two parameters, as shown in Fig. 11.1.

**D. Summary**

The $LD_{50}$ and $LC_{50}$ of MIK were determined in mice after oral, intraperitoneal, and inhalational administration. Results showed an oral $LD_{50}$ of 1.90 ± 0.68 g/kg, and an intraperitoneal $LD_{50}$ of 0.59 ± 0.23 g/kg. By inhalation for 45 min, MIK has an $LC_{50}$ of 74.2 ± 25.8 mg/l. Comparison of the $LD_{50}$ and $LC_{50}$ of MIK and methyl ethyl ketone proved that, by all routes of administration, MIK is more toxic than the other ketone. The ratio of the $LD_{50}$ of MIK to that of methyl ethyl ketone, calculated on a molecular basis, is 43.7% by the oral route and 25.7% by intraperitoneal injection. The $LC_{50}$ showed a value of 26.1%. A correlation was found in this study between the molar refractivity and acute toxicity of various ketones and alcohols.

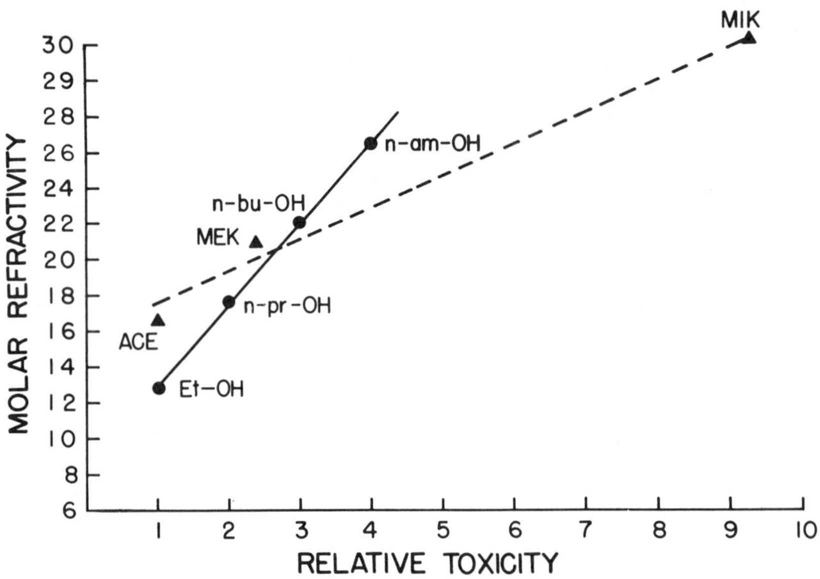


FIGURE 11.1. The correlation between the toxicities of ketones and alcohols, and their molar refractivity calculated according to Wulf and Featherstone.[50] ACE = acetone, MEK = methyl ethyl ketone, MIK = methyl isobutyl ketone, EtOH = ethanol, *n*-pr-OH = *n*-propanol, *n*-bu-OH = *n*-butyl alcohol, *n*-am-OH = *n*-amyl alcohol.

Chapter 12

# PULMONARY AND SYSTEMIC VASCULAR RESPONSE TO METHYL ISOBUTYL KETONE IN THE CAT

As noted in the review section of this monograph (Chapter 10), the authors are unaware of any investigations concerning the hemodynamic responses following MIK administration. In light of a previous finding of this laboratory that methyl ethyl ketone has potent pulmonary vasoconstrictor properties in the open-chest cat and dog (Chapters 8 and 9), it was appropriate to carry out a similar investigation for MIK. Results of experiments in which cats inhaled MIK are reported in this chapter.

## A. Methods

Nine male cats, weighing 2.9 to 3.4 kg (mean ± SE, 3.2 ± 0.06 kg) were anesthetized with pentobarbital sodium, 30 mg/kg, by intraperitoneal injection. The trachea, left femoral artery, and left femoral vein were cannulated. Blood pressure was monitored by a Statham P23AA transducer connected to the arterial cannula, and recorded on a Sanborn 7700 recorder. Prior to opening the chest, mechanical ventilation (20 breaths/min, 30 to 50 ml/breath) was begun with a Harvard small animal respirator. After these procedures, the thorax was opened by an incision in the fifth left intercostal space. The small pulmonary artery to the left lower lobe was cannulated (PE 50), and recording of pulmonary arterial pressure begun using a low-pressure transducer (Statham P23BB). Left atrial pressure was monitored in a similar manner, following insertion of a cannula (PE 60) into the left atrial appendage. Pulmonary arterial blood flow, also representing cardiac output, was monitored electronically by a Statham noncannulating flow probe (6 to 8 mm diameter), model SP 7515, placed around the main pulmonary artery and connected to a Statham SP 2202 flow meter. Following these procedures, the animal was permitted to stabilize (about 30 min), during which time the lungs were periodically inflated to prevent atelectasis.

Concentrations of MIK were prepared for inhalation by volatilizing a measured volume of MIK liquid into a stream of air entering a breathing bag at a set rate. Following a control period of breathing with room air, the bag was connected to the inlet of the respirator. The cat was exposed to MIK vapor, 5 min/concentration, according to a sequence of increasing concentrations (v/v): 0.01% (0.41 mg/l), 0.05% (2.05 mg/l), 0.10% (4.1 mg/l), 0.25% (10.25 mg/l), 0.5% (20.5 mg/l) and 1.0% (41.0 mg/l). Additionally, in some experiments, MIK was also injected intravenously in geometrically increasing doses (4 to 128 mg/l), in order to determine (a) if pulmonary effects were limited to the inhalational route of administration, and (b) the difference between the lethal dose and the threshold dose for a pulmonary vascular response.

## B. Calculations and Abbreviations of Circulatory Measurements

MPAP: Mean pulmonary arterial pressure in cm $H_2O$, measured from the pulmonary artery to the left lower lobe.

MLAP: Mean left atrial pressure in cm $H_2O$, measured from the left atrium.

EMPAP: Effective mean pulmonary arterial pressure in cm $H_2O$, mean pulmonary arterial pressure minus mean left atrial pressure.

MAP: Mean arterial pressure in mm Hg, measured from the left femoral artery.

MPAF: Mean pulmonary arterial flow in ml/min, measured with a Statham electromagnetic flow probe around the main pulmonary artery, using a Statham electromagnetic flowmeter.

PVR: Pulmonary vascular resistance in dynes·sec/$cm^5$, the quotient of mean pulmonary arterial pressure minus mean left atrial pressure in dynes/$cm^2$, and of mean pulmonary arterial flow in ml/sec.

SVR: Systemic vascular resistance in dynes·sec/$cm^5$, the quotient of mean arterial pressure in dynes/$cm^2$ and of mean pulmonary arterial flow in ml/sec.

The paired t-test was used to determine the statistical significance of mean differences in values from control and experimental trials. Differences in P values of 0.05 and less were considered statistically significant.

## C. Results

Portions of the record from an inhalation experiment with MIK are shown in Figure 12.1. As previously seen with methyl ethyl ketone (Chapter 8), the most pronounced and consistent effect of

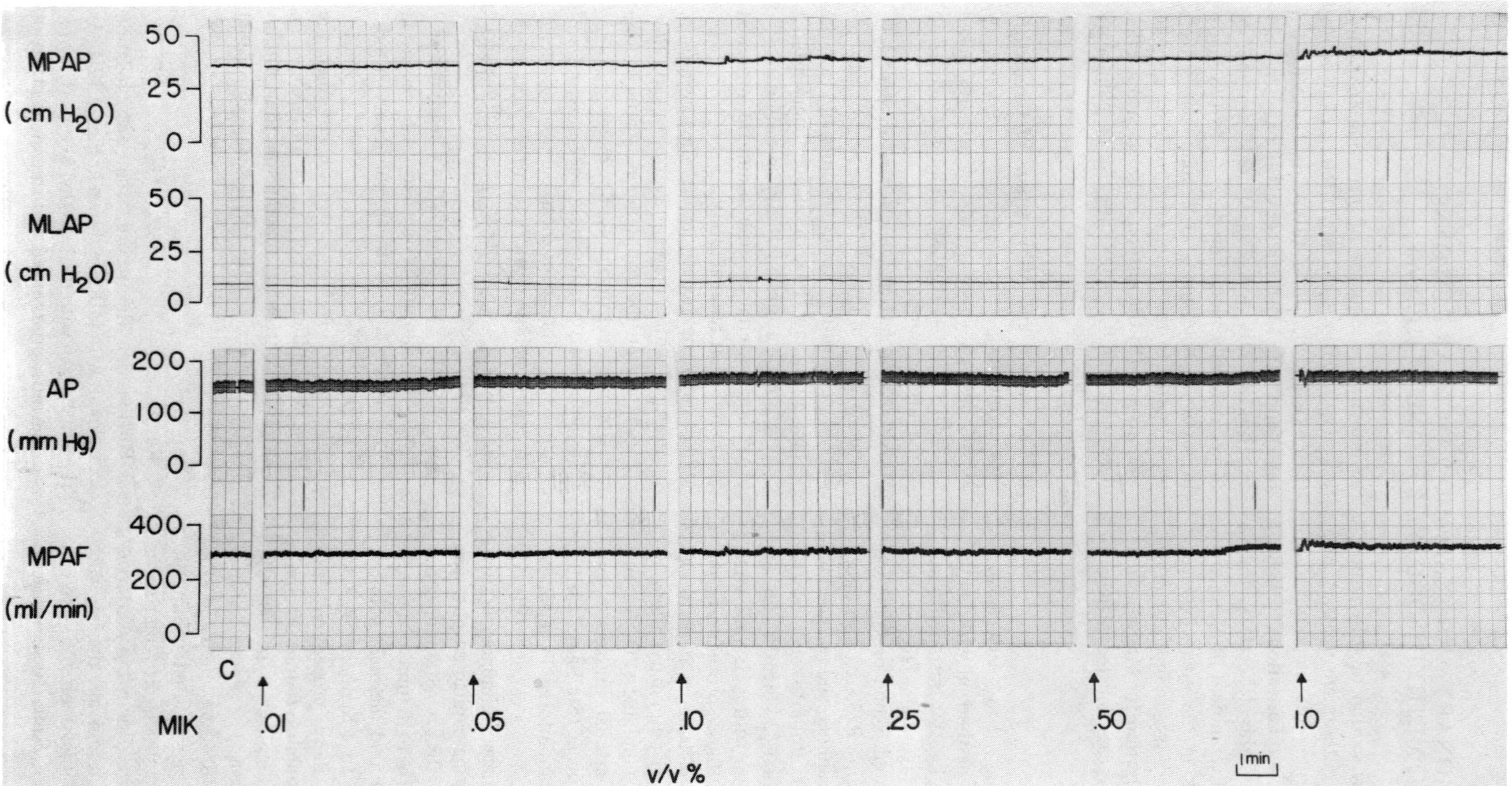


FIGURE 12.1. The effect of progressively increasing concentrations of MIK on some circulatory parameters in the open-chest cat. The record shown includes MPAP: mean pulmonary arterial pressure; MLAP: mean left atrial pressure; AP: arterial pressure; MPAF: mean pulmonary arterial flow. Note that an increase in MPAP appears at the lowest concentration tested (0.01%).

MIK on measured parameters is a concentration-dependent increase over controls in mean pulmonary arterial pressure. Figure 12.1 shows that these elevations initially appeared during inhalation of the lowest concentration, 0.01%. Changes in mean arterial blood flow usually represented a trend toward recovery following an initial significant decline. As for systemic effects of MIK, changes in mean arterial pressure were variable, with no overall pattern seen. Mean left atrial pressure (not shown in Figure 12.1) was virtually unchanged by inhalation of MIK.

A summary of effects of MIK inhalation on pulmonary and systemic circulatory factors, both measured and derived, is given for nine experiments in Table 12.1. Figure 12.2 represents log concentration-response curves showing the percent change from control averages in these parameters. Inhalation of MIK in all concentrations was found to cause pulmonary hypertension and vasoconstriction, but virtually no change in systemic hemodynamic parameters.

*1. Pulmonary Vasoconstriction*

As can be seen in Figure 12.2, increases in both mean pulmonary arterial pressure and pulmonary vascular resistance, in contrast to normal or below-normal mean pulmonary arterial flow, demonstrate that all concentrations of MIK induced pulmonary vasoconstriction. Inhalation of 0.01% elicited a small but significant increase in mean pulmonary arterial pressure (2%, $P < 0.05$). This increase, since it was during exposure to the lowest concentration tested, prevented determination of the threshold concentration. Greater elevation in pulmonary vascular resistance, 7% ($P < 0.01$), calculated from the ratio of effective mean pulmonary arterial pressure and mean pulmonary arterial flow, was due to a statistically significant reduction from controls in pulmonary arterial blood flow (-3%, $P < 0.05$). Both mean pulmonary arterial pressure and pulmonary vascular resistance continued to increase with increasing MIK concentrations, greatest elevations (9% and 18%, respectively) occurring during inhalation of 1.0% MIK. These increases were accompanied by a relatively steady recovery in mean pulmonary arterial flow to control levels.

*2. Systemic Vascular Resistance*

In contrast to consistent, statistically significant effects on the pulmonary vasculature, changes in systemic factors during MIK inhalation were small and inconsistent. Nonsignificant increases of 3 to 4% occurred in mean arterial pressure and systemic vascular resistance during inhalation of concentrations ranging from 0.01% to 0.25% MIK. Although the administration of 0.5% MIK was virtually without effect, 1.0% MIK induced a nonsignificant 4 to 5% decline from control averages in both mean arterial pressure and systemic vascular resistance.

*3. Effect of Intravenous Administration on Pulmonary Vascular Responses to MIK*

Effects of MIK on the pulmonary vascular bed were not confined to the inhalational route. Pulmonary vasoconstriction and hypertension also occurred following intravenous administration of MIK. Results of a representative experiment are shown in Figure 12.3. Administration of MIK, 4 mg/kg, elicited no response. Following the injection of MIK, 8 mg/kg, there was an elevation in mean pulmonary arterial pressure (17%) and a decrease in mean pulmonary arterial flow (-34%). That a substantial increase in pressure took place despite a large decline in blood flow indicates an intense increase in pulmonary vascular resistance (84%). This increase in resistance was most likely caused by pulmonary vasoconstriction rather than passively by a decline in cardiac function (shown by the precipitous fall in cardiac output and, briefly, systemic arterial pressure). This conclusion is based on the facts that (a) measurements were made during the plateau response after recovery of mean arterial pressure, and (b) subsequent higher doses of MIK brought about further elevations in mean pulmonary arterial pressure without further depression of cardiac output. For purposes of comparison with bronchopulmonary effects of MIK described in the next section (Chapter 13), it might be noted that pulmonary vasoconstriction occurred following intravenous injection of only 8 mg/kg, while the lethal dose of MIK by this route was between 64 and 128 mg/kg.

**D. Discussion**

Based on a comparison of the threshold concentrations, MIK administered by the inhalational route was found to be a more potent pulmonary vasoconstrictor than methyl ethyl ketone in the open-chest cats (the subject of Chapter 8). While the latter substance induced significant increases in mean pulmonary arterial pressure and pulmonary

TABLE 12.1

**Effect of Increasing Concentrations of MIK on Circulatory Dynamics in the Cat***

| | MPAP cm $H_2O$ | MLAP cm $H_2O$ | EMPAP cm $H_2O$ | MAP mmHg | MPAF ml/min | Vascular resistances dynes·sec/cm$^5$ Pulmonary | Vascular resistances dynes·sec/cm$^5$ Systemic |
|---|---|---|---|---|---|---|---|
| Control | 31.3 ± 1.26 | 6.8 ± 0.84 | 26.5 ± 1.64 | 126.5 ± 7.65 | 334.1 ± 20.93 | 4564.8 ± 336.55 | 32483.1 ± 2762.89 |
| MIK | 32.0 ± 1.29 | 6.7 ± 0.82 | 27.7 ± 1.66 | 128.4 ± 7.94 | 322.8 ± 18.98 | 4860.5 ± 323.27 | 33011.0 ± 3064.55 |
| 0.01% | 0.7 ± 0.27 | −0.1 ± 0.15 | 1.2 ± 0.22 | 1.9 ± 1.48 | −11.4 ± 3.84 | 295.7 ± 51.16 | 527.9 ± 1078.34 |
| | 0.05 | NS | 0.01 | NS | 0.05 | 0.01 | NS |
| MIK | 31.8 ± 1.02 | 7.1 ± 0.81 | 26.7 ± 1.29 | 127.6 ± 7.24 | 328.1 ± 23.42 | 4877.2 ± 339.80 | 33850.5 ± 3073.11 |
| 0.05% | 1.3 ± 0.35 | 0.4 ± 0.24 | 1.2 ± 0.38 | 1.1 ± 1.34 | −6.6 ± 4.77 | 312.3 ± 106.69 | 1367.4 ± 618.73 |
| | 0.01 | NS | 0.05 | NS | NS | 0.05 | NS |
| MIK | 33.2 ± 1.24 | 7.0 ± 0.82 | 28.1 ± 1.34 | 129.0 ± 7.64 | 329.6 ± 19.58 | 5035.0 ± 341.44 | 33855.9 ± 3342.87 |
| 0.10% | 3.0 ± 0.38 | 0.2 ± 0.20 | 1.6 ± 0.52 | 2.5 ± 2.48 | −4.5 ± 8.01 | 470.2 ± 134.01 | 1372.8 ± 881.41 |
| | 0.01 | NS | 0.05 | NS | NS | 0.02 | NS |
| MIK | 33.1 ± 1.21 | 6.9 ± 0.77 | 28.3 ± 1.39 | 127.6 ± 7.18 | 327.3 ± 22.43 | 5116.5 ± 389.69 | 33367.3 ± 3077.06 |
| 0.25% | 1.8 ± 0.52 | 0.1 ± 0.28 | 1.9 ± 0.75 | 1.1 ± 2.28 | −6.9 ± 9.30 | 551.7 ± 206.89 | 884.1 ± 524.16 |
| | 0.02 | NS | NS | NS | NS | 0.05 | NS |
| MIK | 33.6 ± 1.38 | 7.0 ± 0.83 | 28.9 ± 1.59 | 126.9 ± 8.01 | 334.3 ± 24.44 | 5124.0 ± 427.63 | 32492.4 ± 3096.93 |
| 0.50% | 2.3 ± 0.68 | 0.2 ± 0.28 | 2.4 ± 0.86 | 0.4 ± 3.62 | 0.1 ± 12.41 | 559.2 ± 255.00 | 9.3 ± 573.77 |
| | 0.02 | NS | 0.05 | NS | NS | NS | NS |
| MIK | 34.0 ± 1.63 | 7.1 ± 0.85 | 29.6 ± 1.83 | 119.4 ± 8.90 | 336.1 ± 25.73 | 5339.8 ± 415.00 | 31634.9 ± 3381.65 |
| 1.0% | 2.7 ± 1.03 | 0.3 ± 0.31 | 3.1 ± 1.22 | −7.1 ± 6.29 | 2.0 ± 16.30 | 775.0 ± 220.99 | 848.3 ± 863.72 |
| | 0.05 | NS | NS | NS | NS | 0.02 | NS |

*Data are mean ± SE of control and experimental values, and the mean ± SE of the difference between these values, summarized from nine experiments. The t-test for paired data was used to determine statistical significance of the mean differences. MPAP: mean pulmonary arterial pressure; MLAP: mean left atrial pressure; EMPAP: effective mean pulmonary arterial pressure; MAP: mean arterial pressure; MPAF: mean pulmonary arterial flow.

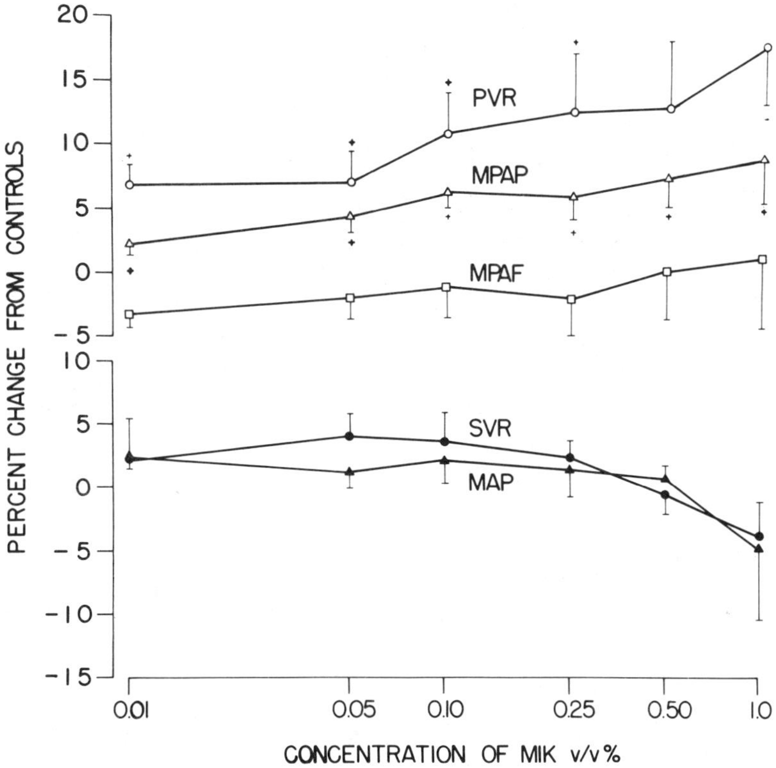


FIGURE 12.2. Mean percentage changes from control averages in response to progressively increasing concentrations of MIK in nine cats. PVR: pulmonary vascular resistance; MPAP: mean pulmonary arterial pressure; MPAF: mean pulmonary arterial flow; SVR: systemic vascular resistance; MAP: mean arterial pressure. Effective mean pulmonary arterial pressure, which responded similarly as MPAP, is not shown. Bars represent the SE of the mean. + = statistically significant mean difference from control averages, according to the t-test for paired data ($P \leqslant 0.05$).

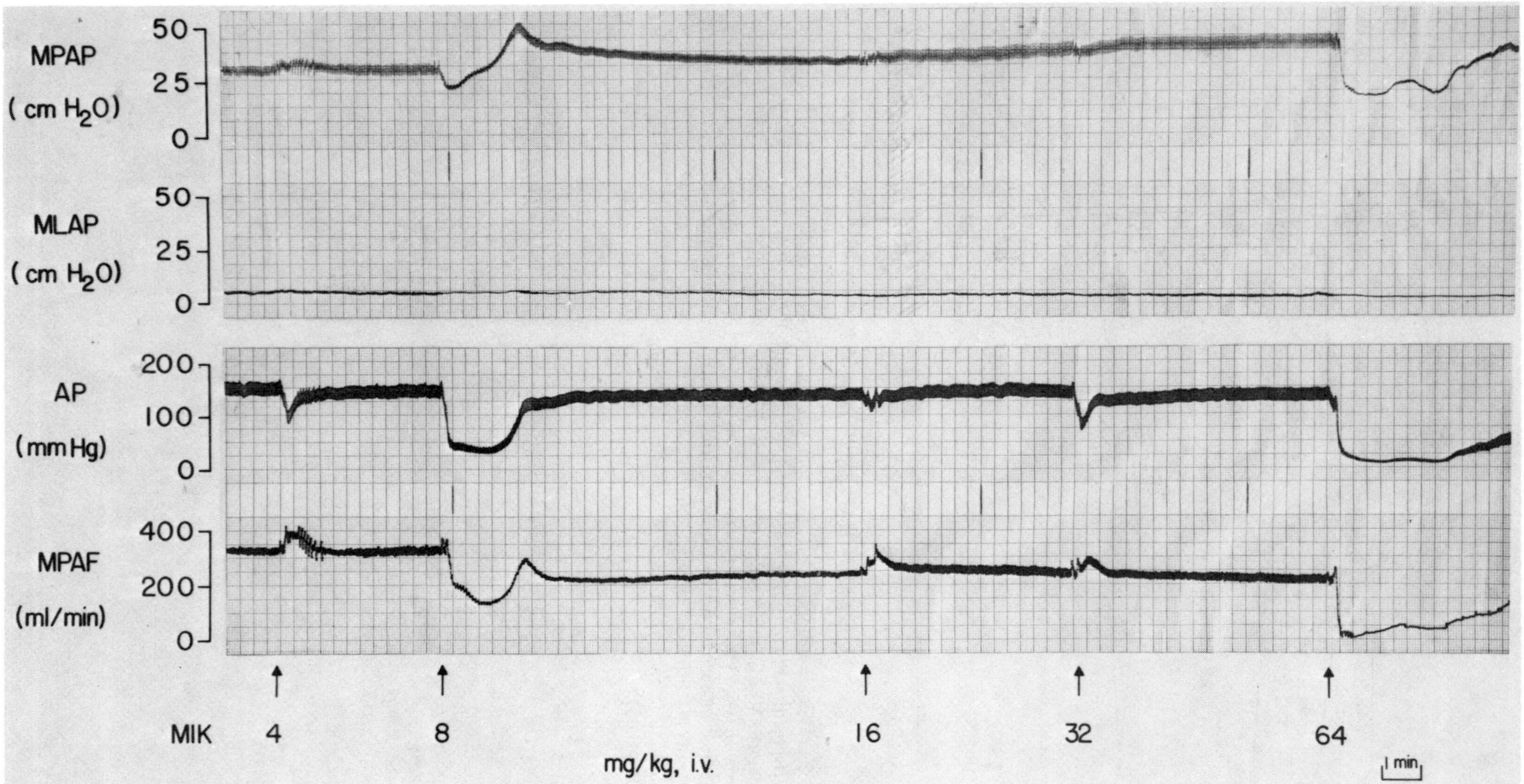


FIGURE 12.3. The effect of intravenous administration of increasing doses of MIK in the open-chest cat. The record shown includes MPAP: mean pulmonary arterial pressure; MLAP: mean left atrial pressure; AP: arterial pressure; MPAF: mean pulmonary arterial flow. The continued increase in MPAP, without further decline in MPAF, indicates pulmonary vasoconstriction (see text).

vascular resistance during inhalation of 0.10% and 0.05% respectively, significant changes in both these parameters were noted during inhalation of only 0.01% MIK. This difference is even greater on a molar basis, since the molecular weight of MIK is greater than that of methyl ethyl ketone. In light of the current TLV[28] for MIK (0.01% or 0.41 mg/l), it is thus possible for workers and others to be exposed to levels of MIK capable of causing pulmonary vasoconstriction and hypertension.

Although the manner by which MIK elicits pulmonary hypertension and vasoconstriction is unknown, previous experiments with methyl ethyl ketone suggest a complex mode of action, possibly including histamine and catecholamine release as well as stimulation of the sympathetic innervation to the lungs.

**E. Summary**

Experiments in the open-chest cat demonstrated significant pulmonary hypertension and vasoconstriction, as well as reduction in mean pulmonary arterial flow, during inhalation of the lowest MIK concentration tested (0.01% or 0.41 mg/l). Effects on the pulmonary vasculature were further augmented with increasing concentrations of MIK, while blood flow slowly recovered. In contrast to these effects on the lungs, there was virtually no change in systemic hemodynamic factors (mean arterial pressure and systemic vascular resistance) except for a small, nonsignificant decline during inhalation of the highest concentration (1.0% or 41 mg/l).

Similar findings of pulmonary hypertension and vasoconstriction following intravenous administration of MIK (threshold dose 8 mg/kg), indicated that pulmonary vascular effects of MIK are not limited to administration by the inhalational route. The sensitivity of pulmonary vascular smooth muscle to MIK was illustrated by the much larger dose (64 to 128 mg/kg) required for lethal effects.

These findings, along with the results of a previous study in this laboratory (Chapter 8), mark the second commonly used ketone causing pulmonary hypertension in the cat.

Chapter 13

# BRONCHOPULMONARY RESPONSES TO METHYL ISOBUTYL KETONE IN THE CAT

In light of evidence from this laboratory that MIK (and previously, methyl ethyl ketone) causes pulmonary vasoconstriction and pulmonary hypertension in the cat (Chapter 8 and 12), experiments have been carried out to determine if MIK-induced stimulation in the lung is specific for vascular smooth muscle or affects bronchopulmonary smooth muscle as well. This chapter describes studies in closed-chest cats, demonstrating the effect of MIK on pulmonary (airway) resistance and other factors associated with respiratory mechanics. As in the preceding hemodynamic study (Chapter 12), some animals were also given MIK intravenously, in order to determine (a) if MIK-induced bronchoconstriction is limited to administration by the inhalational route, and (b) the difference between the lethal dose and the threshold dose for increases in airway resistance. An unexpected finding, the appearance of a Bezold-Jarisch reflex-like effect following intravenous administration of MIK, is also discussed.

## A. Methods

Respiratory responses during exposure to MIK vapor were studied in free-breathing, closed-chest cats. Eight male cats, weighing from 2.7 to 3.3 kg (mean ± SE, 3.1 ± 0.08 kg) were anesthetized by intraperitoneal injection of α-chloralose (50 mg/kg) and pentobarbital sodium (10 mg/kg). The trachea, left femoral artery, and left femoral vein were cannulated. Blood pressure was monitored by a Statham P23 Gb transducer connected to the femoral arterial cannula, and recorded on the blood pressure channel of a Pulmonary Mechanics Computer chart-writer (Buxco Electronics, Sharon, Connecticut).

Respiratory parameters were electronically derived by the analog computer from two basic input signals: flow rate and transpulmonary pressure. Air flow was monitored by a heated Fleish pneumotachograph in series with the tracheal cannula, with differences across the screen measured by a Statham model 11282 differential pressure transducer. Transpulmonary pressure was measured as the difference between intrapleural pressure and tracheal pressure, using a Validyne MP 45-1 differential pressure transducer. For this purpose a cannula attached to one side of the transducer was inserted into the pleural cavity at the fifth or sixth right intercostal space, while the other side of the transducer was connected to a sidearm in the tracheal cannula. The following measured and computed parameters were monitored: air flow rate, transpulmonary pressure, tidal volume, dynamic compliance, pulmonary (airway) resistance, respiratory rate, and minute volume. Tidal volume, in milliliters, was integrated from the expiratory flow rate (ml/sec) for each breath, and multiplied by the respiratory rate (breaths/min) to give minute volume (ml/min). Pulmonary resistance, indicating bronchomotor tone, was computed in cm $H_2O$/l/min, by dividing differential transpulmonary pressures and flow rates ($\Delta P/\Delta F$), sampled at isovolumetric levels in the inspiratory and expiratory sides of the respiratory cycle. Dynamic compliance, measuring the stretching ability or ease of lung expansion, was computed in ml/cm $H_2O$, by dividing volume and pressure sampled at the two points of zero flow corresponding to maximal and minimal tidal volumes.

For determining respiratory effects of MIK inhalation, vapor concentrations were prepared in breathing bags as described in Chapter 12 and connected to an inlet port leading to the tracheal cannula. A similar port, opposite in position, allowed expiration into a waste collecting bag. A constant stream of vapor from which the animal could breathe at will was provided by a pump placed between the vapor bag and the inlet port. Following a period of control ventilation with a bag containing charcoal-filtered air, the animal was exposed for 5-min durations to each of the following concentrations of MIK (v/v): 0.01%, 0.05%, 0.10%, 0.25%, 0.5%, and 1.0%.

As in the experiments in Chapter 12 measuring circulatory effects of MIK, some animals were also given intravenous injections of MIK, ranging from 4 to 128 mg/kg.

## B. Results

A representative experiment showing the effect of MIK inhalation on bronchopulmonary function in the cat is presented in Figure 13.1. It can be seen that overall, both pulmonary (airway) resistance and transpulmonary pressure increased from controls, elevations beginning during the final 3

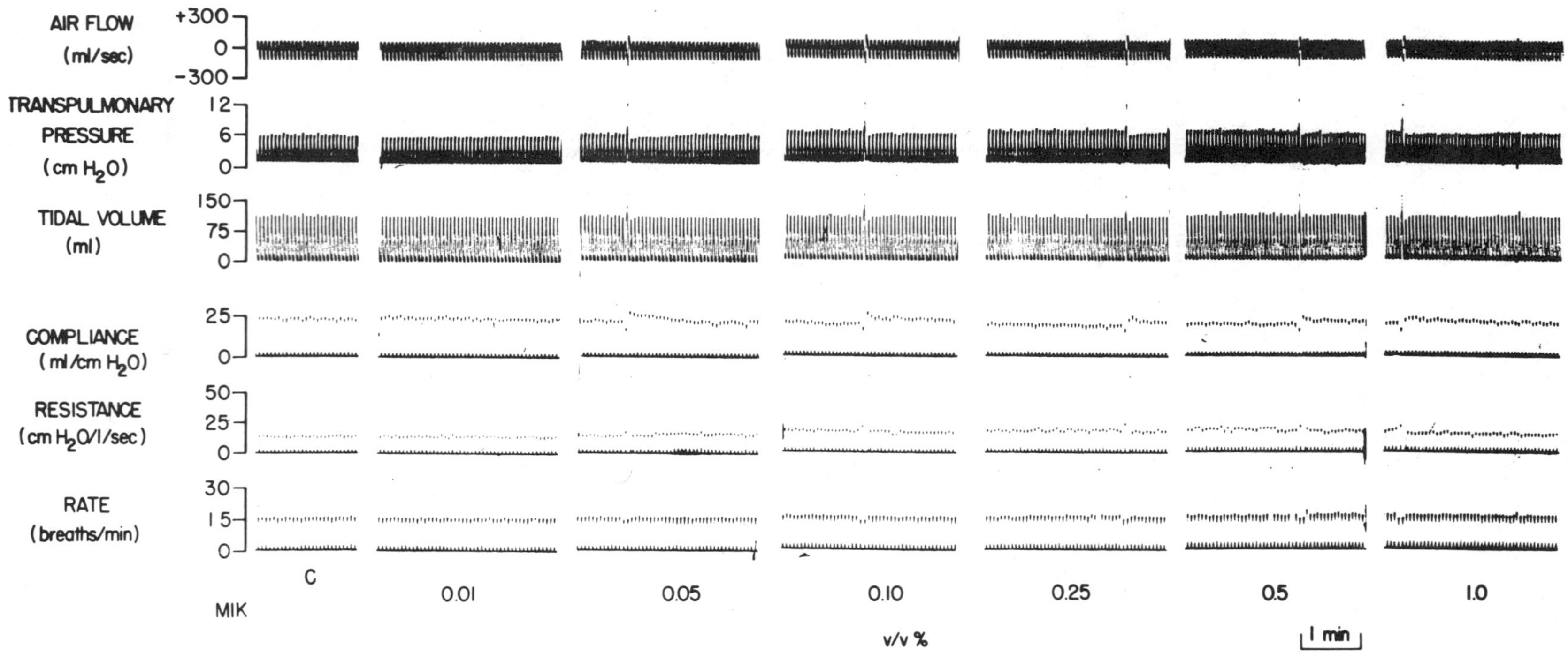


FIGURE 13.1. Effect of increasing concentrations of MIK on respiration in the closed-chest cat. All parameters are computed from air flow (inspiration down) and transpulmonary pressure (negative up). Each panel represents the final 3 min of a 5-min trial.

min of ventilation with 0.05% MIK and reaching a maximum during inhalation of 0.5% MIK. An overall trend of decreased dynamic compliance can also be noted, beginning with inhalation of 0.05% and attaining a maximum decrease during ventilation with 0.25% MIK. The magnitude of these bronchopulmonary responses was somewhat attenuated during inhalation of 1.0% MIK.

Data summarized from eight experiments are presented in Table 13.1. In Figure 13.2 are log concentration-response curves, illustrating percent changes from control averages in parameters of major significance to this study. The primary finding from inhalation experiments was bronchoconstriction, based on a small but statistically significant elevation in pulmonary resistance without increase in tracheal air flow.

*1. Bronchoconstriction*

As can be seen in Figure 13.2, transpulmonary pressure and pulmonary resistance were similarly elevated from controls in response to MIK inhalation. These increases represented an elevation in bronchomotor tone due to bronchoconstriction, since flow rate was virtually unaltered. While small average increases in pressure and resistance (3% and 4%, respectively) appeared during inhalation of 0.01% MIK, the first statistically significant elevation in resistance (7%, $p \leqslant 0.05$) was found with 0.05% and in pressure (5%, $p \leqslant 0.02$) with 0.10% MIK. Concentrations of 0.10% and 0.25% MIK elicited further elevations in pulmonary resistance (8% and 9%) and in transpulmonary pressure (8% with 0.25% MIK, nonsignificant). Pressure alone continued to increase (8%) during ventilation with 0.5% MIK, while 1.0% MIK caused a further increase in resistance (11%) but attenuation of the pressure response (5%).

*2. Effects on Compliance and Respiration*

Effects on dynamic compliance were small and variable, with a maximum decrease (−3%) during inhalation of 0.25% MIK. Recovery of compliance appeared during ventilation with 0.5% and 1.0% MIK. In contrast, tidal volume, which was slightly elevated (3% to 4%) by concentrations of 0.05% to 0.25% MIK, was even further increased (still statistically nonsignificant) by 0.5% MIK (6%) and 1.0% MIK (8%). The larger increase in tidal volume coincided with the recovery of dynamic compliance.

*3. Effect of Intravenous Administration of MIK: the Bezold-Jarisch Reflex*

MIK administered by intravenous injection did not elevate pulmonary resistance at any dose level. Figure 13.3, illustrating a representative experiment, shows that intravenous injection of doses ranging from 4 to 32 mg/kg did not increase pulmonary resistance or transpulmonary pressure. Unexpectedly, following the injection of 64 mg/kg, there was a simultaneous precipitous hypotension and apneic response, similar to effects seen previously by Aviado and co-workers following intravenous injection of veratridine in dogs, and representing the Bezold-Jarisch reflex.[44] This effect was even more pronounced in other experiments, with apneic periods as long as 45 seconds prior to resumption of breathing.

The inability of intravenously administered MIK to elevate pulmonary resistance, seen repeatedly in various experiments, indicates that MIK-induced bronchoconstriction may be limited to the inhalational route.

**C. Discussion**

These experiments represent the first reported demonstration of bronchoconstriction during exposure to MIK vapor, and of an apnea-hypotension response, similar to the Bezold-Jarisch reflex, following intravenous administration. The only previous indications of the pneumotoxicity of MIK in laboratory animals were the findings by Specht[5] of decreased respiratory rate in guinea pigs exposed to 1.68% (68.9 mg/l), and currently, of pulmonary vasoconstriction and pulmonary hypertension in cats (Chapter 9) and dogs (Chapter 14).

Of potential significance to elucidating the manner in which MIK elicits bronchoconstriction is the report by Elkins[27] that respiratory tract irritation was a primary complaint of workers in one of two cases in the United States of adverse reactions during exposure to the TLV concentration of MIK (100 ppm, or 0.41 mg/l). The relevance of this observation is that stimulation of irritant receptors in the upper and lower respiratory tract, either by mechanical means or by inhalation of irritants (sulfur dioxide, inert dusts), has been demonstrated by Nadel and co-workers[45,46] and Widdicombe et al.[47] to cause bronchoconstriction. This action has been attributed to vagal reflexes, since it is abolished following

TABLE 13.1

**Effect of Increasing Concentrations of MIK on Bronchopulmonary Function in the Cat***

| | Air flow ml/sec | Transpulmonary pressure cm/$H_2O$ | Tidal volume ml | Compliance ml/cm $H_2O$ | Resistance cm $H_2O$/l/sec | Rate breaths/min | Minute volume ml/min |
|---|---|---|---|---|---|---|---|
| Control | 202.6 ± 21.73 | 5.5 ± 0.60 | 84.4 ± 6.22 | 16.9 ± 1.42 | 13.0 ± 1.36 | 15.0 ± 1.30 | 1250.4 ± 160.59 |
| MIK 0.01% | 204.6 ± 22.27<br>2.0 ± 2.58<br>NS | 5.6 ± 0.65<br>0.1 ± 0.09<br>NS | 83.3 ± 6.26<br>0.9 ± 1.16<br>NS | 16.8 ± 1.41<br>−0.1 ± 0.23<br>NS | 13.6 ± 1.57<br>0.6 ± 0.29<br>NS | 14.9 ± 1.36<br>−0.1 ± 0.13<br>NS | 1262.4 ± 171.87<br>12.0 ± 19.44<br>NS |
| MIK 0.05% | 204.6 ± 22.27<br>2.0 ± 2.58<br>NS | 5.8 ± 0.66<br>0.3 ± 0.12<br>NS | 84.6 ± 6.29<br>2.1 ± 1.45<br>NS | 16.4 ± 1.09<br>−0.4 ± 0.43<br>NS | 13.9 ± 1.53<br>0.9 ± 0.34<br>0.05 | 14.9 ± 1.34<br>−0.1 ± 0.17<br>NS | 1287.7 ± 180.63<br>37.3 ± 30.57<br>NS |
| MIK 0.10% | 204.9 ± 23.49<br>2.3 ± 4.26<br>NS | 5.8 ± 0.64<br>0.3 ± 0.09<br>0.02 | 85.6 ± 6.78<br>3.1 ± 2.33<br>NS | 16.4 ± 1.06<br>−0.5 ± 0.46<br>NS | 14.1 ± 1.62<br>1.1 ± 0.40<br>0.05 | 14.9 ± 1.39<br>−0.1 ± 0.21<br>NS | 1305.1 ± 196.07<br>54.7 ± 48.69<br>NS |
| MIK 0.25% | 205.4 ± 23.18<br>2.9 ± 4.25<br>NS | 5.9 ± 0.63<br>0.4 ± 0.18<br>NS | 84.9 ± 6.25<br>2.6 ± 2.95<br>NS | 16.1 ± 0.95<br>−0.8 ± 0.54<br>NS | 14.3 ± 1.64<br>1.3 ± 0.42<br>0.05 | 14.7 ± 1.42<br>−0.3 ± 0.24<br>NS | 1289.1 ± 195.97<br>38.7 ± 53.33<br>NS |
| MIK 0.50% | 203.4 ± 21.42<br>0.9 ± 5.05<br>NS | 5.9 ± 0.63<br>0.4 ± 0.15<br>0.05 | 87.0 ± 6.65<br>4.6 ± 2.54<br>NS | 16.2 ± 1.01<br>−0.6 ± 0.46<br>NS | 14.1 ± 1.67<br>1.1 ± 0.46<br>0.05 | 14.7 ± 1.67<br>−0.3 ± 0.38<br>NS | 1319.6 ± 204.22<br>68.9 ± 56.04<br>NS |
| MIK 1.0% | 202.9 ± 21.76<br>0.2 ± 5.10<br>NS | 5.8 ± 0.66<br>0.3 ± 0.12<br>0.05 | 89.0 ± 7.08<br>6.6 ± 3.68<br>NS | 16.6 ± 0.91<br>−0.2 ± 0.70<br>NS | 14.4 ± 1.61<br>1.4 ± 0.53<br>0.05 | 14.8 ± 1.49<br>−0.2 ± 0.42<br>NS | 1360.9 ± 214.29<br>110.4 ± 83.01<br>NS |

*Data shown are the mean ± SE of control and experimental values, and beneath, the mean ± SE of the difference between control and experimental values, summarized from eight experiments. The t-test was used to determine statistical significance of the mean differences.

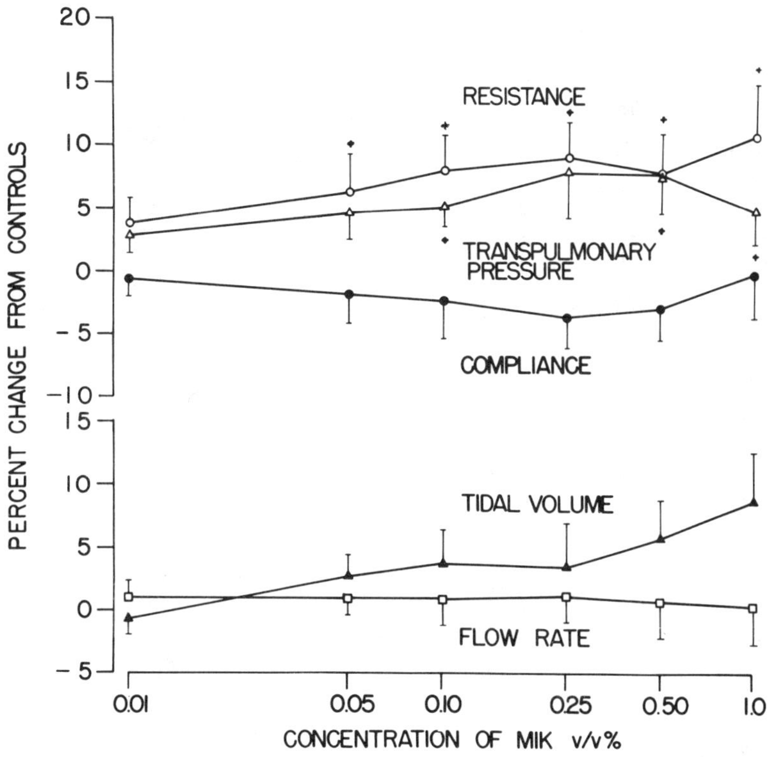


FIGURE 13.2. Percent change from control averages in respiratory parameters during inhalation of increasing concentrations of MIK. Results of eight experiments are summarized. + = statistically significant mean difference from control averages, using the t-test for paired data ($P \leqslant 0.05$).

vagotomy. Such a reflex mechanism may also be important in the response to inhaled MIK, which is highly irritating to a variety of mucous membranes in animals and man (see Chapter 10). Although this hypothesis is appealing due to its simplicity, some complicating factors exist. Nadel, finding that in man atropine sulfate blocked the bronchoconstriction due to irritants, postulated a possible postganglionic cholinergic pathway as well. Additionally, histamine release may contribute to the bronchoconstriction following MIK. A previous report from this laboratory demonstrated that diphenhydramine in antihistaminic doses can at least partially block the pulmonary hypertension and vasoconstriction during inhalation of a similar substance, methyl ethyl ketone (see Chapter 8). These problems remain to be resolved.

While pulmonary vascular responses have been demonstrated following MIK administration via either the inhalational or intravenous routes, only inhalational administration of MIK elicited bronchoconstriction. Instead, there was apnea and hypotension following high intravenous doses (32 to 64 mg/kg), which resembled the Bezold-Jarisch reflex response noted by Aviado and co-workers[44] in dogs injected intravenously with veratridine, a veratrum alkaloid. If MIK has a similar action as veratridine, experiments by Aviado and associates[44] indicate that bilateral vagotomy would nullify all effects. Although the response to intravenous MIK appears to mimic that to veratridine, there could be some differences. Aviado and associates[44] localized the receptors for the hemodynamic effects to the heart. Further localization to vagal afferents supplying this area has been recently shown by Juhasz-Nagy and Szentivanyi.[48] Experiments with veratridine aerosol indicated receptors for apnea in the peripheral portions of the lung, in communication with both the circulation and the alveolar spaces.[44] In the current experiments, apnea did not appear following inhalation of MIK. While this difference may likely be due to a subthreshold concentration reaching the alveoli, it might also suggest a complex action for MIK: the resultant effect of

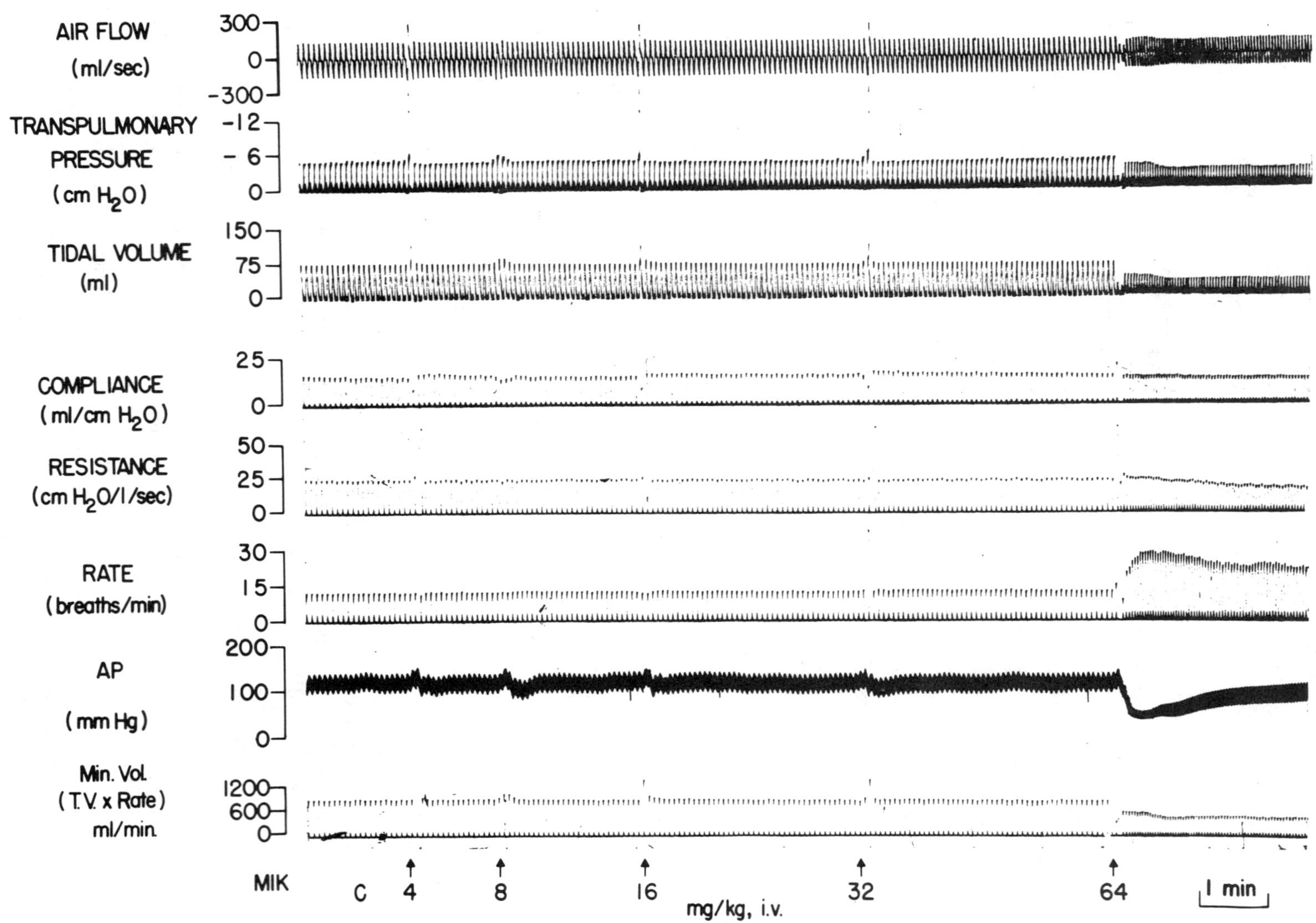


FIGURE 13.3. Effect of intravenous administration of MIK on blood pressure and respiratory factors. Note apnea accompanied by severe hypotension following administration of 64 mg/kg (Bezold-Jarisch reflex-like effect). Minute volume before reflex effect, 1050 ml/min, and after, 860 ml/min.

stimulation of respiratory tract irritant receptors described by Nadel and others[45-47] as well as receptors eliciting the apnea associated with veratradine and the Bezold-Jarisch reflex found by Aviado and co-workers.[44]

### D. Summary

The current experiments demonstrate that MIK inhalation induces limited but significant bronchoconstriction as well as pulmonary vasoconstriction and pulmonary hypertension in the cat. In contrast, intravenous administration of MIK, while causing pulmonary vasoconstriction, does not elicit bronchoconstriction. Since the inhalational route is, therefore, necessary for bronchoconstriction, and MIK is known to irritate mucous membranes as found in the respiratory tract, it is possible that MIK-induced bronchoconstriction is by stimulation of respiratory tract irritant receptors, and mediated by vagal reflexes.

Both apnea and severe hypotension are seen following intravenous administration of near-lethal doses of MIK. These effects appear similar to the Bezold-Jarisch reflex response following intravenous injection of veratridine in dogs. Therefore, it is possible that intravenous administration of MIK, as in the case with veratridine, stimulates pulmonary receptors in the peripheral portions of the lung to cause apnea, and receptors in the heart that initiate the precipitous fall in blood pressure. Since the Bezold-Jarisch reflex did not occur during inhalation of MIK, and the bronchoconstriction that did occur was limited, it is possible that MIK administered by this route may have a complex mechanism, involving both irritant receptors in the respiratory tract (bronchoconstriction) and pulmonary receptors for apnea.

Chapter 14

# MYOCARDIAL AND HEMODYNAMIC EFFECTS OF METHYL ISOBUTYL KETONE IN THE DOG

In an earlier report from this laboratory, methyl ethyl ketone was reported to induce pulmonary hypertension in dogs and cats (Chapters 8 and 9). Since methyl isobutyl ketone (MIK), a higher homologue of methyl ethyl ketone, is extensively used in industry and in various consumer products (Chapter 10), it appeared necessary to investigate its effects on the pulmonary and systemic circulation. The pharmacodynamic effects of MIK administered by inhalation are described in this chapter.

## A. Methods

Adult mongrel dogs of either sex, weighing between 18 and 24 kg (average 20.6 kg), were anesthetized with 30 to 35 mg/kg pentobarbital sodium intravenously. The chest was opened and various hemodynamic parameters studied, as previously described.[53] The following concentrations of MIK were administered, via the inlet of the respirator, for 5 min: 0.01%, 0.05%, 0.10%, 0.25%, 0.50%, and 1.0%. The minimal effective concentration of MIK was determined and then administered to the dog preparation either alone or mixed with the threshold concentration of trichlorofluoromethane (FC 11) previously determined in a similar preparation. Statistical analysis of the data followed the same criteria used with methyl ethyl ketone (Chapter 9).

## B. Calculations and Abbreviations of Hemodynamic Measurements

Mean pulmonary arterial pressure (MPAP), mean left atrial pressure (MLAP), left ventricular pressure (LVP), left ventricular end diastolic pressure (LVEDP), left ventricular dp/dt, mean aortic pressure (MAP), and mean pulmonary arterial flow (MPAF) were measured as previously described. Effective mean pulmonary arterial pressure (EMPAP), heart rate (HR), systemic and pulmonary vascular resistances, stroke volume, and stroke work were calculated as previously mentioned (Chapter 9).

## C. Results

A summary of the hemodynamic effects of MIK in the anesthetized dog preparation is shown in Table 14.1. Figure 14.1 shows a record of a typical experiment with MIK. The results of a mixture of the threshold concentration of MIK alone, or mixed with the threshold concentration of trichlorofluoromethane (FC 11), are shown in Table 14.2.

**Increase in pulmonary arterial pressure and pulmonary vascular resistance** – In concentrations as low as 0.05%, MIK produced an increase of 5.1%, 5.8%, and 6.9% in mean pulmonary arterial pressure, effective mean pulmonary arterial pressure, and pulmonary vascular resistance, respectively. The rise in these parameters is intensified by increase in the concentration of inhaled MIK, reaching a maximum of 8.3%, 12.1%, and 12.2% for the above-mentioned parameters with 0.25% MIK. Increasing MIK concentration beyond 0.25% resulted in a nonsignificant increase in the same parameters (Figure 14.2).

**Decrease in myocardial contractility** – MIK in concentrations of 0.01% and 0.05% was without effect on the inotropic state of the heart muscle as reflected by changes in maximum rate of rise of left ventricular pressure (dp/dt). Increasing the concentration of inhaled MIK to 0.1%, 0.25%, and 0.5% brought about a nonsignificant decrease in myocardial contractility. Further increase of MIK concentration to 1%, however, elicited a significant decrease of 15.9% in myocardial contractility as compared with the control value. At this concentration, viz., 1%, a significant decrease in left ventricular pressure of 11.3% and a significant decrease in systemic vascular resistance of 10.3% were observed.

**Effect on heart rate and arterial pressure** – A consistent and concentration-dependent (although nonsignificant) increase in heart rate was observed after inhalation of various concentrations of MIK. Thus, 0.05%, 0.10%, and 0.25% MIK are associated with an increase of 0.8%, 1.1%, and 2.8% in the heart rate, respectively. The same concentrations brought about a nonsignificant increase in mean aortic pressure of 1.8%, 1.5%, and 1.5%, respectively. On the other hand, inhalation of 0.5% and 1.0% MIK brought about an increase in heart rate of 2.6% and 0.7% accompanied by a decrease in mean aortic pressure of 1.3% and 10.6%, respectively.

**Effect on cardiac output, stroke volume and**

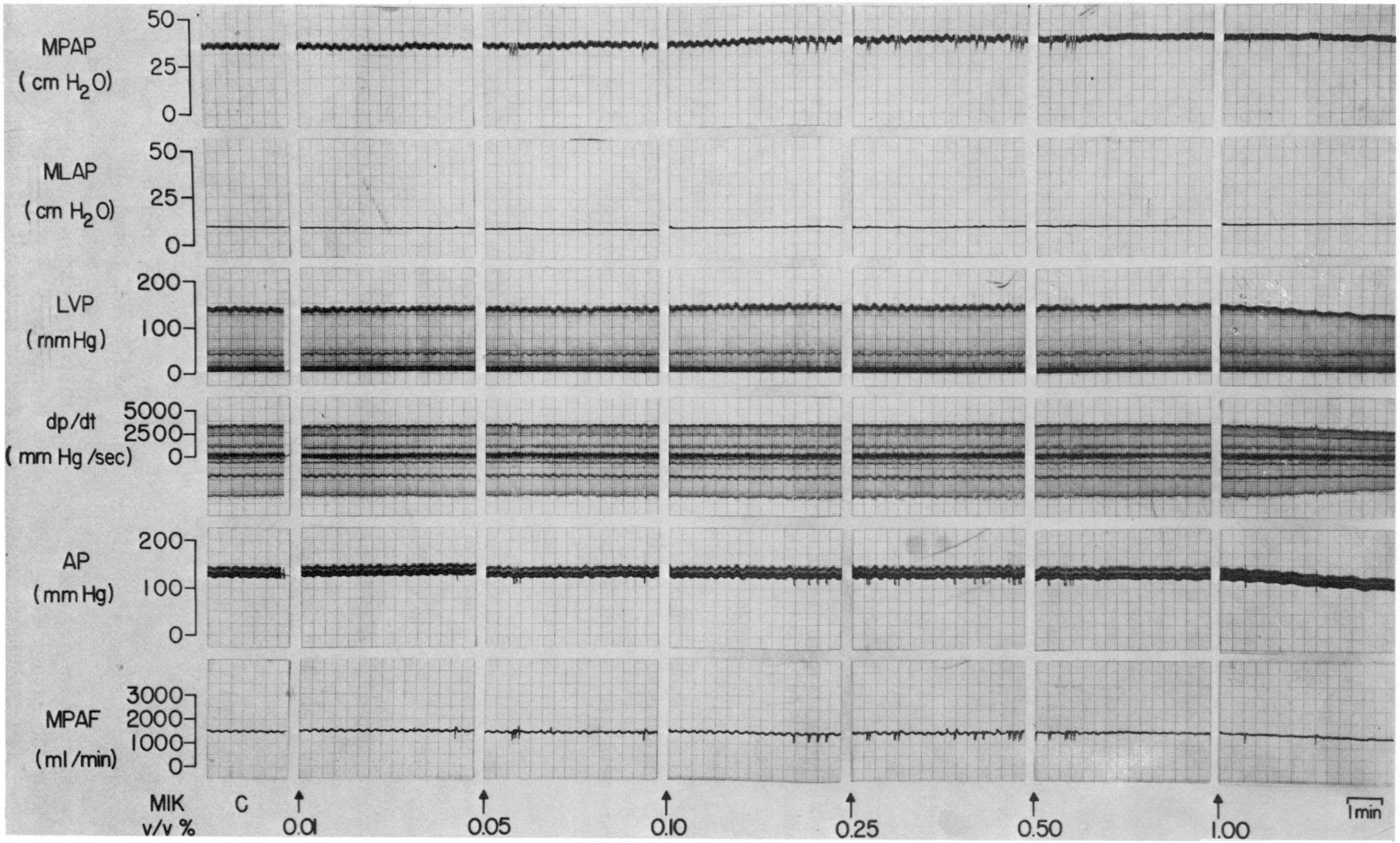


FIGURE 14.1. The effect of progressively increasing (0.01 to 1.0%) concentrations of MIK on the cardiovascular system of the open-chest dog preparation. Note the increase in pulmonary arterial pressure with various concentrations and the myocardial depression induced by 1% of MIK.

TABLE 14.1

**The Effect of Progressively Increasing Concentrations of MIK on the Hemodynamics of the Open-chest Dog***

| | MPAP cm $H_2O$ | MLAP cm $H_2O$ | EMPAP cm $H_2O$ | LVP mmHg | LVEDP mmHg | dp/dt mmHg/sec | MAP mmHg | MPAF ml/min | HR beats/min | Vascular resistance dynes·sec/cm$^5$ | | Stroke vol ml | Stroke work g·m |
|---|---|---|---|---|---|---|---|---|---|---|---|---|---|
| | | | | | | | | | | Pulmonary | Systemic | | |
| Control | 28.2 ± 3.57 | 6.8 ± 0.86 | 21.4 ± 3.40 | 126 ± 6.14 | 1.2 ± 0.73 | 3958 ± 307.3 | 105 ± 8.97 | 1525 ± 79.3 | 177 ± 7.89 | 864 ± 168.4 | 5582 ± 638.5 | 8.7 ± 0.64 | 12.9 ± 1.29 |
| 0.01% MIK | 29.4 ± 3.89<br>+0.6 ± 0.27<br>NS | 7.7 ± 1.10<br>+0.2 ± 0.15<br>NS | 21.7 ± 3.61<br>+0.5 ± 0.16<br>NS | 125 ± 9.72<br>−1.3 ± 1.3<br>NS | 1.2 ± 0.73<br>0 ± 0.77<br>NS | 3750 ± 2700<br>0 ± 51.03<br>NS | 111 ± 11.13<br>+1.3 ± 1.49<br>NS | 1525 ± 82.9<br>+12.5 ± 12.50<br>NS | 179 ± 13.1<br>+0.3 ± 1.03<br>NS | 861 ± 166.8<br>+15.3 ± 5.91<br>NS | 5873 ± 818.8<br>+99.8 ± 59.8<br>NS | 8.7 ± 0.89<br>+0.1 ± 0.11<br>NS | 13.6 ± 1.67<br>+0.1 ± 0.32<br>NS |
| 0.05% MIK | 31.8 ± 3.60<br>+1.5 ± 0.26<br>0.01 | 7.1 ± 1.00<br>+0.1 ± 0.12<br>NS | 24.6 ± 3.70<br>+1.4 ± 0.30<br>0.02 | 129 ± 1.03<br>−0.4 ± 1.3<br>NS | 1.6 ± 1.03<br>+0.4 ± 1.02<br>NS | 4050 ± 421.3<br>0 ± 68.47<br>NS | 110 ± 8.51<br>+1.4 ± 2.44<br>NS | 1460 ± 91.4<br>−10 ± 29.15<br>NS | 180 ± 10.6<br>+1.6 ± 1.69<br>NS | 1034 ± 188.9<br>+74.6 ± 19.39<br>0.02 | 6126 ± 694.7<br>+181.8 ± 132.17<br>NS | 8.3 ± 0.97<br>−0.1 ± 0.22<br>NS | 12.9 ± 1.24<br>+0.1 ± 0.55<br>NS |
| 0.10% MIK | 32.1 ± 3.98<br>+1.9 ± 0.78<br>0.05 | 7.0 ± 1.02<br>0 ± 0<br>NS | 25.1 ± 3.98<br>+1.9 ± 0.78<br>0.05 | 127 ± 8.58<br>−1.4 ± 2.09<br>NS | 1.8 ± 0.73<br>+0.06 ± 1.12<br>NS | 3900 ± 420.9<br>−150 ± 91.86<br>NS | 110 ± 9.62<br>+1.4 ± 2.44<br>NS | 1470 ± 76.8<br>0 ± 35.36<br>NS | 180 ± 10.0<br>+2.0 ± 1.22<br>NS | 1031 ± 178.9<br>+71.8 ± 19.24<br>0.05 | 6044 ± 682.2<br>+99.6 ± 58.28<br>NS | 8.3 ± 0.71<br>−0.1 ± 0.21<br>NS | 12.9 ± 1.52<br>−0.02 ± 0.58<br>NS |
| 0.25% MIK | 32.9 ± 0.07<br>+2.6 ± 0.72<br>0.05 | 6.9 ± 1.05<br>−0.1 ± 0.14<br>NS | 26.0 ± 3.94<br>+2.8 ± 0.66<br>0.02 | 128 ± 8.08<br>0 ± 1.58<br>NS | 2.2 ± 0.97<br>+1.0 ± 1.38<br>NS | 3900 ± 420.9<br>−150 ± 91.86<br>NS | 110 ± 8.73<br>+1.0 ± 3.73<br>NS | 1470 ± 76.8<br>0 ± 35.36<br>NS | 183 ± 10.7<br>+5.2 ± 1.93<br>NS | 1068 ± 179.3<br>+108.8 ± 19.93<br>0.01 | 6023 ± 6497<br>+78.8 ± 115.76<br>NS | 8.1 ± 0.74<br>−0.2 ± 0.25<br>NS | 12.7 ± 1.46<br>−0.3 ± 0.76<br>NS |
| 0.50% MIK | 32.6 ± 4.18<br>+2.4 ± 0.78<br>NS | 6.6 ± 1.18<br>−0.2 ± 0.24<br>NS | 26.0 ± 4.04<br>+2.6 ± 0.87<br>NS | 126 ± 8.65<br>−2.6 ± 2.29<br>NS | 1.6 ± 1.03<br>+0.4 ± 1.61<br>NS | 3875 ± 3992<br>−150 ± 72.89<br>NS | 106 ± 9.77<br>−1.8 ± 3.92<br>NS | 1480 ± 66.3<br>−10 ± 55.68<br>NS | 183 ± 11.2<br>+4.8 ± 1.91<br>NS | 1054 ± 175.7<br>+94.4 ± 53.46<br>NS | 5783 ± 669.4<br>−88.8 ± 70.32<br>NS | 8.3 ± 0.71<br>−0.2 ± 0.34<br>NS | 12.5 ± 1.52<br>−0.5 ± 0.92<br>NS |
| 1.0% MIK | 32.5 ± 4.01<br>+2.2 ± 0.76<br>NS | 6.9 ± 1.43<br>0 ± 0.68<br>NS | 25.6 ± 3.88<br>+2.2 ± 1.20<br>NS | 114 ± 8.28<br>−14.2 ± 3.84<br>0.05 | 1.6 ± 1.03<br>+0.4 ± 1.61<br>NS | 3425 ± 453.5<br>−600 ± 144.70<br>0.02 | 96 ± 8.02<br>−12.4 ± 6.50<br>NS | 1470 ± 62.45<br>−20 ± 64.42<br>NS | 180 ± 12.3<br>+1.8 ± 2.75<br>NS | 1048 ± 173.3<br>+88.6 ± 63.94<br>NS | 5227 ± 550.3<br>−644.6 ± 185.6<br>0.05 | 8.3 ± 0.59<br>−0.2 ± 0.31<br>NS | 11.3 ± 0.97<br>−1.8 ± 1.24<br>NS |

*Each group of numbers represents mean response ± SE, mean difference from its control ± SE of difference and P value. NS = nonsignificant (i.e., $P > 0.05$). Abbreviations are explained in the text.

TABLE 14.2

**The Effect of MIK, FC 11, and a Mixture of Both on the Open-chest Dog***

| | MPAP cm $H_2O$ | | MLAP cm $H_2O$ | | EMPAP cm $H_2O$ | | LVP mmHg | | LVEDP mmHg | | dp/dt mmHg/sec | | MAP mmHg | |
|---|---|---|---|---|---|---|---|---|---|---|---|---|---|---|
| | C | E | C | E | C | E | C | E | C | E | C | E | C | E |
| MIK | 33.5 | 35.0 | 7.2 ± 7.2 | | 26.3 | 27.8 | 131.5 | 130.8 | 0.8 | 0.8 | 4155 | 4074 | 113.8 | 113.8 |
| | +1.5 ± 0.78 | | 0 ± 0 | | +1.5 ± 0.78 | | −0.8 ± 2.69 | | 0 ± 0 | | −63 ± 108.3 | | 0 ± 3.54 | |
| | NS | | NS | | NS | | NS | | NS | | NS | | NS | |
| FC 11 0.5% | 33.5 | 33.0 | 7.8 | 7.8 | 25.6 | 25.2 | 125.3 | 121.3 | 1.5 | 1.5 | 3563 | 3406 | 109 | 106 |
| | −0.5 ± 0.46 | | 0 ± 0 | | −0.5 ± 0.46 | | −4.0 ± 1.68 | | 0 ± 0 | | −157 ± 31.3 | | −3.3 ± 1.7 | |
| | NS | | NS | | NS | | NS | | NS | | 0.02 | | NS | |
| MIK (Threshold) +0.5% FC 11 | 33.6 | 33.8 | 7.1 | 6.9 | 26.6 | 26.9 | 127 | 119 | 2.8 | 2.8 | 3625 | 3219 | 112 | 107 |
| | +0.2 ± 0.31 | | −0.2 ± 0.38 | | +0.4 ± 0.39 | | −8.0 ± 2.38 | | 0 ± 0 | | −406 ± 106.7 | | −5.5 ± 2.6 | |
| | NS | | NS | | NS | | 0.05 | | NS | | 0.05 | | NS | |

| | MPAF ml/min | | HR beats/min | | Vascular resistance dynes·sec/cm$^5$ Pulmonary | | Vascular resistance dynes·sec/cm$^5$ Systemic | | Stroke vol ml | | Stroke work g·m | |
|---|---|---|---|---|---|---|---|---|---|---|---|---|
| | C | E | C | E | C | E | C | E | C | E | C | E |
| MIK | 1413 | 1388 | 179 | 181 | 1102 | 1176 | 6393 | 6463 | 8.0 | 7.8 | 13.1 | 12.7 |
| | −25 ± 43.3 | | +2 ± 1.89 | | +74 ± 9.88 | | +65 ± 119.3 | | −0.2 ± 0.28 | | −0.4 ± 0.83 | |
| | NS | | NS | | 0.01 | | NS | | NS | | NS | |
| FC 11 0.5% | 1375 | 1350 | 179 | 180 | 1087 | 1087 | 6225 | 6144 | 7.8 | 7.6 | 12.3 | 11.6 |
| | −25 ± 14.4 | | +0.8 ± 1.93 | | 0 ± 23.4 | | −81 ± 49.0 | | −0.2 ± 0.15 | | −0.8 ± 0.31 | |
| | NS | | NS | | NS | | NS | | NS | | NS | |
| MIK (Threshold) +0.5% FC 11 | 1425 | 1350 | 177 | 180 | 1094 | 1168 | 6194 | 6131 | 8.2 | 7.7 | 13.2 | 11.7 |
| | −75 ± 25.0 | | +3.3 ± 2.29 | | +74.3 ± 13.1 | | −13 ± 158.7 | | −0.6 ± 0.18 | | −1.5 ± 0.44 | |
| | NS | | NS | | 0.02 | | NS | | NS | | 0.05 | |

*Each set of numbers consists of mean control (C), mean experimental (E), mean difference, standard error of difference, and the P value. NS = nonsignificant (i.e., $P > 0.05$). Abbreviations are explained in the text.

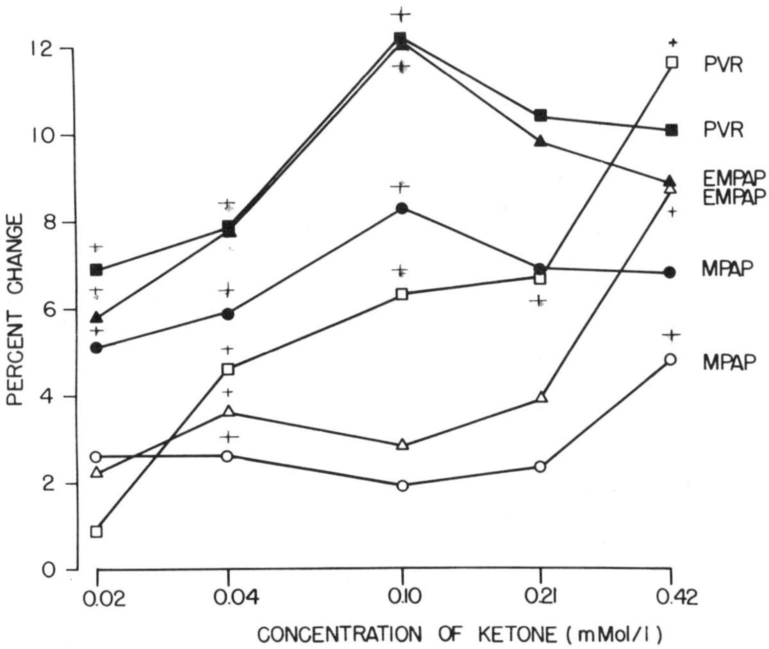


FIGURE 14.2. The effects of various concentrations (mMol/l) of methyl ethyl ketone (○,△,□) and MIK (●,▲,■) on the pulmonary circulation of the open-chest dog preparation. Asterisk shows significant changes. Bars representing SE are omitted for simplicity. Note the greater potency of MIK than methyl ethyl ketone in inducing increase in pulmonary arterial pressure. Abbreviations are explained in the text.

**stroke work** – At high concentrations, namely, 0.5% and 1.0%, MIK brought about a decrease in mean pulmonary arterial flow, stroke volume, and stroke work. This decrease was not statistically significant.

**Effect of MIK and FC 11** – The effect of the threshold concentration of MIK, alone or mixed with the threshold concentration of trichlorofluoromethane (FC 11) was studied in the open-chest dog preparation. The minimal effective concentration of MIK ranged from 0.05% to 0.10% in different preparations with an average of 0.08 ± 0.01%. At this concentration, MIK brought about an increase in mean pulmonary arterial pressure, effective mean pulmonary arterial pressure, and pulmonary vascular resistance, averaging to 4.4%, 5.3%, and 7.0%, respectively; and decreased myocardial contractility by 2.2% (Table 14.2). On the other hand, inhalation of trichlorofluoromethane (FC 11) decreased mean pulmonary arterial pressure, effective mean pulmonary arterial pressure, and mean pulmonary arterial flow by 1.3%, 1.5%, and 1.8%, respectively. Pulmonary vascular resistance, nonetheless, was increased by 0.4%. Both MIK and FC 11 decreased left ventricular pressure by 0.9% and 2.9%, respectively.

Inhalation of a mixture of threshold concentrations of MIK and FC 11 resulted in an increase in mean pulmonary arterial pressure, effective mean pulmonary arterial pressure, and pulmonary vascular resistance by 0.5%, 1.5%, and 7.0%, respectively. The most dramatic effect observed after inhalation of this mixture was on the left ventricular pressure, left ventricular dp/dt, pulmonary arterial flow, and stroke work. A decrease of 6.2%, 10.7%, 5.1%, and 10.8% was brought about in the above-mentioned parameters, respectively. The potentiation of the myocardial depressant effect of FC 11 by MIK is graphically depicted in Figure 14.3b.

**Comparative effects of MEK and MIK** – In a previous study from this laboratory, methyl ethyl ketone was found to bring about, in a similar open-chest dog preparation, an increase in mean pulmonary arterial pressure, effective mean pulmonary arterial pressure, and pulmonary vascular resistance. Since MIK has qualitatively the same effect on the pulmonary vascular bed, it was, therefore, decided to compare the effect of both

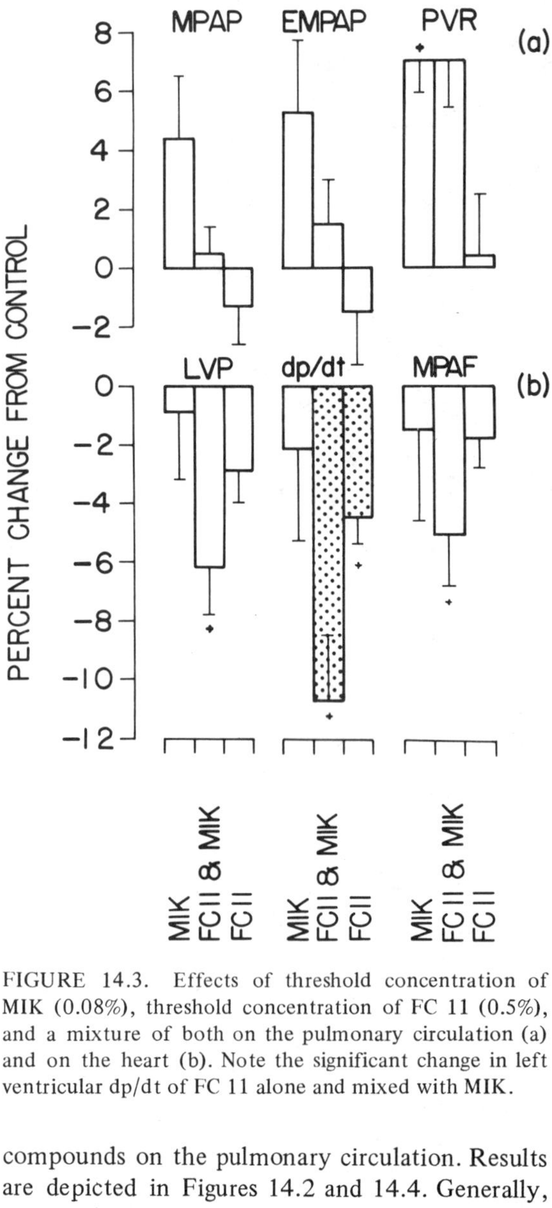


FIGURE 14.3. Effects of threshold concentration of MIK (0.08%), threshold concentration of FC 11 (0.5%), and a mixture of both on the pulmonary circulation (a) and on the heart (b). Note the significant change in left ventricular dp/dt of FC 11 alone and mixed with MIK.

compounds on the pulmonary circulation. Results are depicted in Figures 14.2 and 14.4. Generally, both qualitative and quantitative differences are revealed by this study. At equimolar concentrations, MIK has a greater effect than methyl ethyl ketone on the pulmonary vasculature. Thus, 0.02, 0.04, 0.10, 0.21, and 0.42, mMol/l concentrations are associated with an increase in pulmonary arterial pressure of 2.6, 2.6, 1.9, 2.3, and 4.8% in the case of methyl ethyl ketone and 5.1, 5.9, 8.3, 6.9, and 6.8% in the cases of MIK, respectively. Furthermore, there is a biphasic response in the case of MIK but not in the other ketone. Figure 14.4 shows the effects of 0.1 and 0.42 mmol/l of both ketones. These concentrations are selected on the basis of giving peak effects of MIK and methyl ethyl ketone, respectively. MIK not only elicited greater pulmonary hypertension and myocardial depression than methyl ethyl ketone but also appeared to have qualitatively different effects than the other ketone.

**D. Discussion**

**Effects of MIK** – As in the case of methyl ethyl ketone, the most prominent effect of MIK is an increase in pulmonary arterial pressure and pulmonary vascular resistance. Figure 14.3 shows that there is a significant and concentration-dependent increase in mean pulmonary arterial pressure, effective mean pulmonary arterial pressure and pulmonary vascular resistance, induced by concentrations of from 0.05 to 0.25% (0.02 to 0.10 mMol/l). At these concentration levels, there was almost no change in cardiac output. The increase in pulmonary arterial pressure is due, therefore, to an increase in pulmonary vascular resistance. However, this increase in pulmonary vascular resistance declined with further increase in the concentration of MIK in the inhalation mixture. According to Henry's law,[51] the amount of MIK that passes into solution in the blood is proportional to its partial pressure. Consequently, at high concentrations, 0.5% and 1.0%, a larger amount is expected to dissolve in the blood. Furthermore, Ferguson[52] stated that the biological activity of gases is mainly determined by the ratio of Pt/Ps, where Pt is the partial pressure of the gas and Ps is its saturated vapor pressure. In light of these expected relationships, the observed attenuation in rise of the pulmonary vascular resistance is difficult to explain only on the basis of the physicochemical properties that govern the diffusion and solubility of gases. However, this might be due to other toxic effects of MIK that oppose or overshadow its effect on pulmonary vasculature, or it might be due to tachyphylaxis.

The effect of MIK on the myocardial contractility deserves a comment. At a concentration of 1.0%, MIK brought about a 15.9% decrease in myocardial contractility. However, this concentration was without significant effect on either cardiac output (–0.6%), or stroke volume (–1.2%). That cardiac output was reduced only nonsignificantly despite the significant decrease in myocardial contractility is apparently due to a peripheral mechanism, viz., a decrease in afterload and/or an increase in the effective filling pressure as a consequence of increased venous return. This

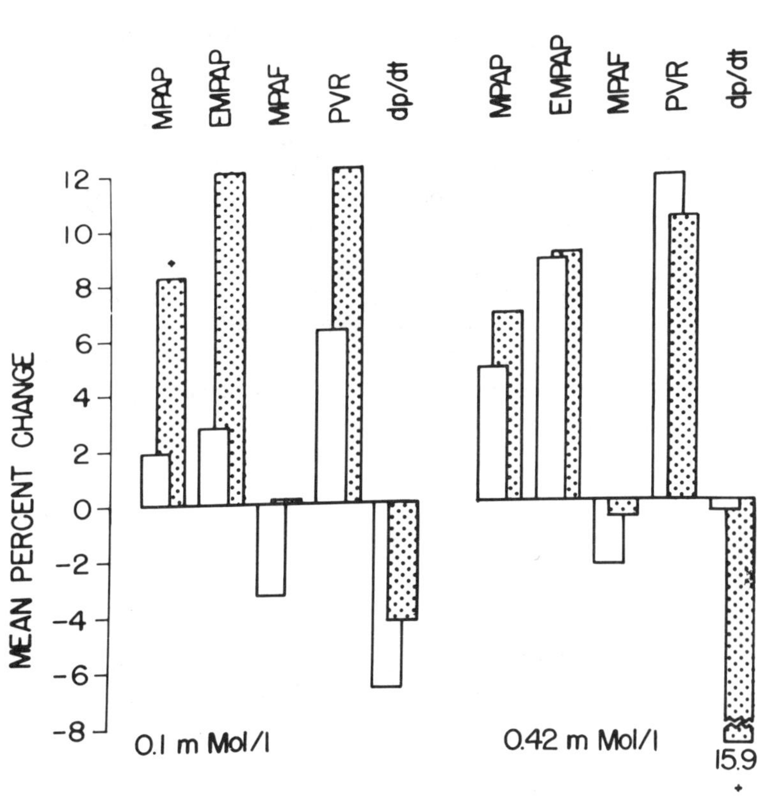


FIGURE 14.4. Comparison of the myocardial and hemodynamic effects of methyl ethyl ketone (open columns) and MIK (stippled columns). Note the significant increase in pulmonary arterial pressure and decrease in dp/dt under the influence of 0.1 and 0.42 mMol/l of MIK, respectively. Bars representing SEM are omitted for simplicity. Asterisk denotes significant changes.

is in line with the finding that, at this concentration, a significant decrease in systemic vascular resistance of 10.3% from the average control was observed. The decrease in left ventricular and mean arterial pressures observed with high concentrations of MIK may be due mainly to decrease in myocardial contractility and either increase in pulmonary vascular resistance or decrease in systemic vascular resistance, respectively.

**Effect of MIK and FC 11** – Figure 14.3 shows that while the threshold concentration of MIK brought about an increase in mean pulmonary arterial pressure and pulmonary vascular resistance, yet threshold concentration of FC 11 exhibited a decrease in the former and a slight increase in the latter parameter. The increase in pulmonary vascular resistance in the case of MIK is due to its effect on the pulmonary vasculature, while in the case of FC 11, the apparent increase in pulmonary vascular resistance is mainly due to greater decrease in pulmonary arterial flow that outweighs the decrease in effective mean pulmonary arterial pressure.

However, both compounds bring about, at the minimal effective concentration, a decrease in myocardial contractility, left ventricular pressure, and mean pulmonary arterial flow. Administration of a mixture of both compounds led to two sets of pharmacologic responses: (a) antagonistic effect on the pulmonary vascular bed that is manifested by a small increase in mean pulmonary arterial pressure, effective mean pulmonary arterial pressure, and nonsignificant increase in pulmonary vascular resistance; and (b) potentiation of the depressant effect on the heart. Thus, a significant decrease in left ventricular pressure, left ventricular dp/dt, and cardiac output was induced by the mixture of both compounds. There is a significant difference between the effect of FC 11 alone and mixed with MIK as indicated by the stippled columns in Figure 14.3b. This possibly implies a different mechanism of action of both compounds.

**Comparison of methyl ethyl ketone and MIK** – Methyl ethyl ketone was previously reported to induce pulmonary hypertension in dogs and cats (Chapters 8 and 9). This study demonstrated that MIK not only produces this effect on dogs as well, but is also more potent than methyl ethyl ketone. Furthermore, MIK in high concentrations exhibits greater affinity to cardiac muscle than methyl ethyl ketone. This is in agreement with results obtained in this laboratory (Chapter 11) comparing the toxicity of both solvents in mice. Since aldehydes are known to be more reactive than ketones, the effect of MIK on the pulmonary vascular bed and on the heart could be due to the ketone itself and/or its metabolic product.

## E. Summary

In the open-chest dog preparation, inhalation of MIK resulted in: (a) pulmonary hypertension, elicited by as low a concentration as 0.05%; and (b) decrease in myocardial contractility only at high concentrations, namely 1%. This myocardial depressant effect was potentiated by inhalation of a mixture of MIK and trichlorofluoromethane (FC 11). Generally, MIK is more potent and more toxic than methyl ethyl ketone.

Chapter 15

# SUMMARY OF THE EFFECTS OF KETONES, CHLORINATED SOLVENTS, FLUOROCARBONS, AND ISOPROPANOL ON THE PULMONARY CIRCULATION

The common feature in the pharmacologic and toxicologic profile of methyl ethyl ketone and methyl isobutyl ketone, uncovered for the first time, is their ability to induce pulmonary hypertension in dogs and cats. In light of the fact that ketones are used extensively in various consumer products either alone or mixed with various solvents and propellants, it was found necessary to summarize the present state of knowledge of the effect of various groups of solvents on the pulmonary circulation. The present investigation (Chapters 4, 9, and 14) and a series of previous ones[53-55] mark the first comprehensive study of these compounds on the pulmonary circulation. Results are summarized in Table 15.1 and Figure 15.1. The compounds can be grouped into two pulmonary hemodynamic patterns.

**Compounds that act primarily on the caliber of large pulmonary vessels** – This group comprises mainly ketones, viz., methyl ethyl ketone and methyl isobutyl ketone. The change in pulmonary arterial blood flow under the influence of these compounds is nonsignificant at any concentration. Furthermore, the threshold concentration for increase in pulmonary arterial pressure and pulmonary vascular resistance is 0.05% in the case of methyl isobutyl ketone and 0.10% in the case of methyl ethyl ketone.

**Compounds that primarily reduce the pulmonary arterial flow** – This group comprises mainly the chlorinated solvents (methyl chloroform and trichloroethylene), the fluorinated propellants (trichlorofluoromethane or FC 11, dichlorodifluoromethane or FC 12, and dichlorotetrafluoroethane or FC 114), and isopropanol. Both chlorinated solvents studied reduce the pulmonary arterial flow at as low a concentration as 0.1%. In this respect, they are more potent than isopropanol, FC 11, and FC 114 which bring about the same effect only at concentrations of 2.5% or greater. FC 12 appears to be the least effective of this group, having a threshold concentration of 10%. The decrease in pulmonary arterial blood flow is accompanied by a depression of the myocardial contractile force.

It is evident from Figure 15.1 that all the compounds that reduce pulmonary blood flow, with the exception of trichloroethylene, bring about a consistent increase in pulmonary vascular resistance (PVR). Since pulmonary vascular resistance is a derived value representing the changes in pulmonary arterial pressure and pulmonary arterial blood flow, it does not, in itself, reflect the state of pulmonary vasomotor activity produced by these compounds. Since pulmonary arterial flow (PAF) is determined by the contractile state of the myocardium, the effect of these compounds on the pulmonary circulation cannot be dissociated from their effect on the myocardium. Decrease in pulmonary arterial flow will lead to an increase in the calculated pulmonary vascular resistance and cause a passive reduction in the caliber of pulmonary blood vessels.[56]

As to the site of action of these solvents and propellants, it is essential to investigate further their mechanism of pulmonary vasoconstriction and myocardial depression. Do these inhalants have a direct effect on the heart and pulmonary vasculature or do they release other substances that cause or contribute to the observed changes? The answer to these questions awaits further research, especially to uncover the possible contribution of pneumotoxicity to death reported from the excessive exposure to these compounds.

TABLE 15.1

**Summary of the Change in Pulmonary Arterial Pressure (cm $H_2O$), Pulmonary Arterial Flow (ml/min) and Pulmonary Vascular Resistance (dynes·sec/$cm^5$)**

(Asterisk denotes the lowest concentration that induces significant change.)

| | Ketones | | Chlorinated solvents | | Fluorinated propellants | | | |
|---|---|---|---|---|---|---|---|---|
| Concentration | Methyl ethyl ketone | Methyl isobutyl ketone | Methyl chloroform | Trichloroethylene | FC 11 | FC 12 | FC 114 | Isopropanol |
| 0.01% | | +0.6 cm $H_2O$<br>+13 ml/min<br>+15 dynes·sec/$cm^5$ | | −2.5 cm $H_2O$<br>0 ml/min<br>−106 dynes·sec/$cm^5$ | | | | |
| 0.05% | +0.8 cm $H_2O$<br>+17 ml/min<br>+5 dynes·sec/$cm^5$ | +1.5 cm $H_2O$*<br>−10 ml/min<br>+75 dynes·sec/$cm^5$ * | +0.7 cm $H_2O$<br>−17 ml/min<br>+26 dynes·sec/$cm^5$ | −2.8 cm $H_2O$<br>−33 ml/min<br>−64 dynes·sec/$cm^5$ | | | | |
| 0.10% | +0.8 cm $H_2O$*<br>−17 ml/min<br>+43 dynes·sec/$cm^5$ * | +1.9 cm $H_2O$<br>0 ml/min<br>+72 dynes·sec/$cm^5$ | −0.2 cm $H_2O$<br>−55 ml/min*<br>+8 dynes·sec/$cm^5$ | +0.4 cm $H_2O$<br>−67 ml/min*<br>+92 dynes·sec/$cm^5$ | | | | |
| 0.25% | +0.7 cm $H_2O$<br>−50 ml/min<br>+63 dynes·sec/$cm^5$ | +2.6 cm $H_2O$<br>0 ml/min<br>+108 dynes·sec/$cm^5$ | +0.4 cm $H_2O$<br>−79 ml/min<br>+72 dynes·sec/$cm^5$ | −1.8 cm $H_2O$<br>−125 ml/min<br>+89 dynes·sec/$cm^5$ | | | | |
| 0.50% | +0.7 cm $H_2O$<br>−33 ml/min<br>+71 dynes·sec/$cm^5$ | +2.4 cm $H_2O$<br>−10 ml/min<br>+94 dynes·sec/$cm^5$ | −1.3 cm $H_2O$<br>−129 ml/min<br>+83 dynes·sec/$cm^5$ | −6.6 cm $H_2O$<br>−133 ml/min<br>−197 dynes·sec/$cm^5$ | −0.6 cm $H_2O$<br>−24 ml/min<br>+179 dynes·sec/$cm^5$ | | | 0 cm $H_2O$<br>−25 ml/min<br>+36 dynes·sec/$cm^5$ |
| 1.00% | +1.6 cm $H_2O$<br>−25 ml/min<br>+131 dynes·sec/$cm^5$ | +2.2 cm $H_2O$<br>−20 ml/min<br>+89 dynes·sec/$cm^5$ | −0.9 cm $H_2O$<br>−171 ml/min<br>+165 dynes·sec/$cm^5$ * | −4.4 cm $H_2O$<br>−275 ml/min<br>−134 dynes·sec/$cm^5$ | +0.2 cm $H_2O$<br>−52 ml/min<br>−112 dynes·sec/$cm^5$ * | | +1.2 cm $H_2O$<br>−17 ml/min<br>+61 dynes·sec/$cm^5$ | −0.4 cm $H_2O$<br>−5 ml/min<br>+49 dynes·sec/$cm^5$ * |
| 2.50% | | | | | −0.8 cm $H_2O$<br>−129 ml/min*<br>+117 dynes·sec/$cm^5$ | | 0 cm $H_2O$<br>−192 ml/min*<br>+134 dynes·sec/$cm^5$ * | −0.2 cm $H_2O$<br>−84 ml/min*<br>+66 dynes·sec/$cm^5$ |
| 5.00% | | | | | −3.3 cm $H_2O$<br>−248 ml/min<br>+514 dynes·sec/$cm^5$ | +0.2 cm $H_2O$<br>−24 ml/min<br>+7 dynes·sec/$cm^5$ | −0.5 cm $H_2O$<br>−100 ml/min<br>+206 dynes·sec/$cm^5$ | −0.1 cm $H_2O$<br>−100 ml/min<br>+47 dynes·sec/$cm^5$ |
| 7.50% | | | | | | | | −1.0 cm $H_2O$<br>−150 ml/min<br>+16 dynes·sec/$cm^5$ |
| 10.0% | | | | | | 0 cm $H_2O$<br>−104 ml/min*<br>+178 dynes·sec/$cm^5$ * | | |
| 20.0% | | | | | −0.6 cm $H_2O$<br>−151 ml/min<br>+212 dynes·sec/$cm^5$ | −1.4 cm $H_2O$<br>−183 ml/min<br>+430 dynes·sec/$cm^5$ | | |

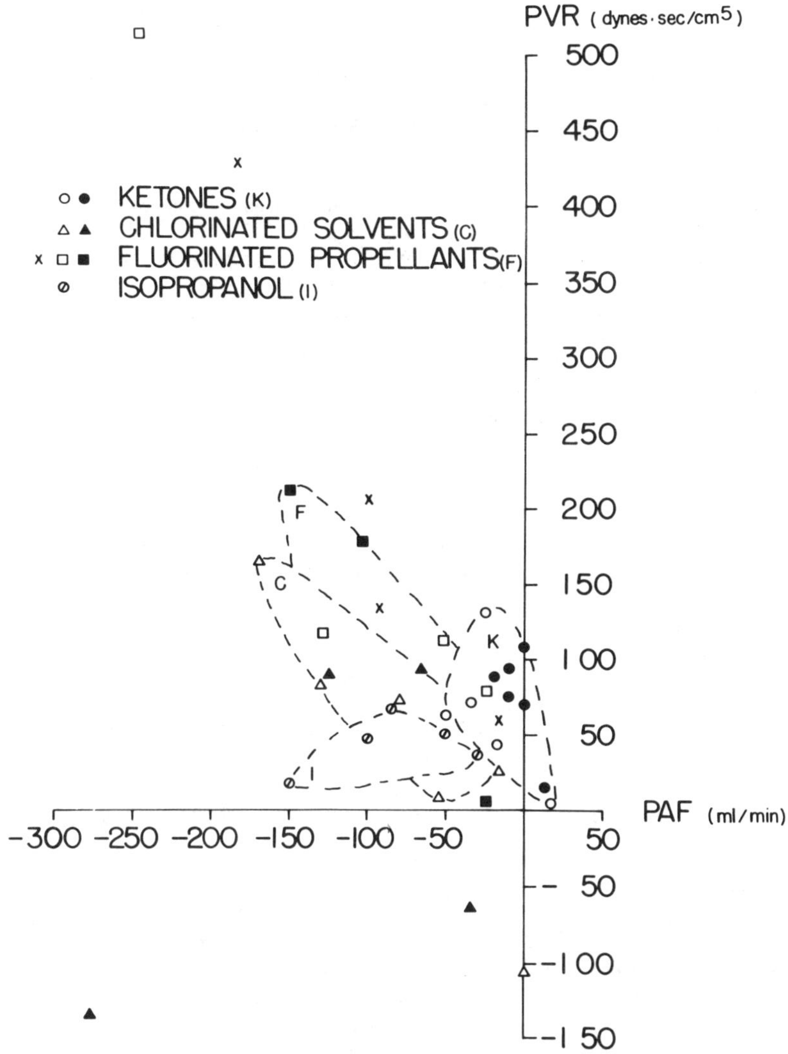


FIGURE 15.1. The effect of various solvents and propellants on the pulmonary circulation. Changes in pulmonary arterial flow (PAF) are plotted against changes in pulmonary vascular resistance (PVR). The relative changes in PAF and PVR of various groups of compounds are shown by corresponding loops.

## REFERENCES – Methyl Isobutyl Ketone


1. **Browning, E.,** *Toxicity and Metabolism of Industrial Solvents,* Elsevier, Amsterdam, 1965.
2. **Weast, R. C., Ed.,** *Handbook of Chemistry and Physics,* 55th ed., CRC Press, Cleveland, 1974.
3. **Jacobs, B.,** *Analytical Chemistry of Industrial Poisons, Hazards, and Solvents,* 2nd ed., Interscience, London, 1949.
4. **Browning, E.,** *Toxicity of Industrial Organic Solvents,* Her Majesty's Stationery Office, London, 1953.
5. **Specht, H.,** Acute response of guinea pigs to inhalation of methyl isobutyl ketone, *U.S. Public Health Rep.,* 53, 292, 1938.
6. **Patty, F. A., Schrenk, H. H., and Yant, W. P.,** Acute responses of guinea pigs to vapors of some new commercial organic compounds. VIII. Butanone, *U.S. Public Health Rep.,* 50, 1217, 1935.
7. **Kovacs, I.,** Lakk- és festékoldószergözök vizsgálata munkahelyek lègterében, *Munkavedelem,* 16, 9, 1970.
8. **Zlatkis, A. and Liebich, H. M.,** Profile of volatile metabolites in human urine, *Clin. Chem.,* 17, 592, 1971.
9. **Purdue, L. J., Enrione, R. E., Thompson, R. J., and Bonfield, B. A.,** Determination of organic and total lead in the atmosphere by atomic absorption spectrometry, *Anal. Chem.,* 45, 527, 1973.
10. **Cholak, J., Yeager, D. W., and Henderson, E. W.,** Determination of lead in biological and related material by atomic absorption spectrophotometry, *Environ. Sci. Technol.,* 5, 1020, 1971.
11. **Blanksma, L. A., Sachs, H. K., Murray, E. F., and O'Connell, M. J.,** Incidence of high blood lead levels in Chicago children, *Pediatrics,* 44, 661, 1969.
12. **Filer, T. D.,** Fluorometric determination of submicrogram quantities of antimony, *Anal. Chem.,* 43, 725, 1971.
13. **Curry, A. S., Read, J. F., and Knott, A. R.,** Determination of thallium in biological material by flame spectrophotometry and atomic absorption, *Analyst* (London), 94, 744, 1969.
14. **Kubasik, N. P. and Volosin, M. T.,** A simplified determination of urinary cadmium, lead, and thallium with use of carbon rod atomization and atomic absorption spectrophotometry, *Clin. Chem.,* 19, 954, 1973.
15. **Gallant, R. W.,** Physical properties of hydrocarbons, Part 27, Ketones, *Hydrocarbon Process.,* 47, 127, 1968.
16. **Sevestre, J.,** Conditions for using ketonic solvents, F.A.T.I.P.E.C. (Fed. Assoc. Techniciens Ind. Peintures, Vernis, Emaux Encres Imprimerie Europe Continentale) 6th, Weisbaden, 1962, 235.
17. **Sevestre, J.,** Ketones for urethane coatings, *Peint. Pigm. Vernis,* 46, 255, 1970.
18. **Dewinter, P. F.,** Methyl isobutyl ketone, solvent for vinyl and acrylic resins, *Chim. Peint.,* 13, 65, 1950.
19. **Bonnat, M.,** Action of solvents on macro-molecular film-formers *Double Liaison,* 112, 41, 1964.
20. **Vivian, T. A.,** Reflowing solvents for repairing surface defects in thermoplastics, German Patent 2,035,999, 1971.
21. **Pino, M. A.,** Oil dewaxing, U.S. Patent 3,320,153, 1967.
22. **Doyle, D. M.,** Removal of metals from lubricating oils and colorimetric analysis for metal content, French Patent 1,460,393, 1966.
23. **Gleason, M. N., Gosselin, R. E., Hodge, H. C., and Smith, R. P.,** *Clinical Toxicology of Commercial Products,* 3rd ed., Williams and Wilkins, Baltimore, 1967.
24. **Nylander, I.,** "Thinner" addiction in children and adolescents, *Acta Paedopsychiatr.,* 29, 273, 1962.
25. **Fujii, M., Shimatani, S., Yoshikawa, T., Tokemoto, I., Kinami, T., and Jitsu, Y.,** *Japan. Kokai,* 6 pp., 1973.
26. **Oda, S. and Yada, H.,** Fish net preservatives with low toxicity to fish, *Japan.,* 4 pp., 1971.
27. **Elkins, B.,** *The Chemistry of Industrial Toxicology,* 2nd ed., John Wiley & Sons, New York, 1959.
28. American Conference of Governmental Industrial Hygienists, Documentation of the Threshold Limit Values for Substances in Workroom Air, 3rd ed., Cincinnati, Ohio, 1971.
29. **Smyth, H. F.,** Hygienic standards for daily inhalation, *Am. Ind. Hyg. Assoc. Q.,* 17, 129, 1956.
30. **Armeli, G., Linari, F., and Martorano, G.,** Rilievi clinici ed ematochimici in operai esposti all'azione di un chetone superiore (MIBK) ripetuti a distanza di 5 anni, *Lav. Um.,* 20, 418, 1968.
31. **Wayne, L. G. and Orcutt, J. A.,** The relative potentials of common organic solvents as precursors of eye-irritants in urban atmospheres, *J. Occup. Med.,* 2, 383, 1960.
32. **May, J.,** Odor thresholds of solvents for evaluating solvent odors in air, *Staub,* 26, 385, 1966.
33. **Silverman, L., Schultz, H. F., and First, M. W.,** Further studies on sensory response to certain industrial solvent vapors, *J. Ind. Hyg. Toxicol.,* 28, 262, 1946.
34. **Smyth, H. F., Carpenter, C. P., and Weil, C. S.,** Range-finding toxicity data: List IV, *Arch. Ind. Hyg. Occup. Med.,* 4, 119, 1951.
35. **Batyrova, T. F.,** Substantiation of the maximum permissible concentration of methyl isobutyl ketone in air of work zones, *Gig. Tr. Prof. Zabol.,* 11, 52, 1973.
36. **Quinby, G. E.,** Solvents as causes of polyneuritis, unpublished paper, Wenatchee, Washington.

36a. **Spencer, P. S. and Thomas, P. K.,** Ultrastructural studies of the dying-back process. II. The sequestration and removal of schwann cells and oligodendrocytes of organelles from normal and diseased axons, *J. Neurocytol.,* in press.

37. Shell Chemical Corp., Safety data sheet, 1957, 57–112.
38. **Specht, H., Miller, J. W., Valaer, P. J., and Sayers, R. R.,** Acute response of guinea pigs to the inhalation of ketone vapors, *U.S. Public Health Serv. Publ.,* National Institute of Health Bulletin No. 176, 1, 1940.
39. **Vernot, E. H., MacEwen, J. D., and Harris, E. S.,** Continuous exposure of animals to methyl isobutyl ketone, *U.S.N.T.I.S., AD Rep.,* Issue No. 751443, 1971.

40. **MacEwen, J. D., Vernot, E. H., and Haun, C. C.,** Effect of 90-day continuous exposure to methylisobutylketone on dogs, monkeys, and rats, *U.S.N.T.I.S. AD Rep.,* Issue No. 730291, 1971.
41. **MacKenzie, W. F.,** Pathological lesions caused by methylisobutylketone, in *Proceedings of the 2nd Annual Conference on Environmental Toxicology,* Aerospace Medical Research Laboratory, Wright-Patterson Air Force Base, Dayton, Ohio, 1971, 311.
42. **Cook, W.,** Maximum allowable concentrations of industrial atmospheric contaminants, *Ind. Med.,* 14, 936, 1945.
43. **Finney, D. J.,** *Probit Analysis,* 3rd ed., Cambridge University Press, 1971.
44. **Aviado, D. M., Pontius, R. G., and Schmidt, C. F.,** The reflex respiratory and circulatory actions of veratridine on pulmonary, cardiac, and carotid receptors, *J. Pharmacol. Exp. Ther.,* 97, 120, 1949.
45. **Nadel, J. A., Salem, H., Ramplin, B., and Tokiwa, Y.,** Mechanism of bronchoconstriction during inhalation of sulfur dioxide, *J. Appl. Physiol.,* 20, 164, 1965.
46. **Nadel, J. A. and Widdicombe, J. G.,** Reflex effects of upper airway irritation on total lung resistance and blood pressure, *J. Appl. Physiol.,* 17, 861, 1962.
47. **Widdicombe, J. G.,Kent, D. C., and Nadel, J. A.,** Mechanism of bronchoconstriction during inhalation of dust, *J. Appl. Physiol.,* 17, 613, 1962.
48. **Juhasz-Nagy, A. and Szentivanyi, M.,** Localization of the receptors of the coronary chemoflex in the dog, *Arch. Int. Pharmacodyn. Ther.,* 131, 39, 1961.
49. **Cammarata, A. and Martin, A. N.,** Physical properties and biological activity, in *Medicinal Chemistry,* 3rd ed., Interscience, New York, 118, 1970.
50. **Wulf, R. J. and Featherstone, R. M.,** A correlation of Van der Waals constants with anesthetic potency, *Anesthesiology,* 18, 97, 1957.
51. **West, J. B.,** *Respiratory Physiology,* Williams and Wilkins, Baltimore, 1974.
52. **Ferguson, J.,** *Proc. Roy. Soc. London Ser. B,* 127, 387, 1939, cited from Lien, E. J., Structure-absorption-distribution relationships, significance for drug design, in *Drug Design,* Vol. 5, Ariens, E. J., Ed., 1975, 82.
53. **Aviado, D. M., Zakhari, S., Simaan, J. A., and Ulsamer, A. G.,** *Methyl Chloroform and Trichloroethylene in the Environment,* CRC Press, 1976.
54. **Simaan, J. A. and Aviado, D. M.,** Hemodynamic effects of aerosol propellants. I. Cardiac depression in the dog, *Toxicology,* 5, 127, 1975.
55. **Simaan, J. A. and Aviado, D. M.,** Hemodynamic effects of aerosol propellants. II. Pulmonary circulation in the dog, *Toxicology,* 5, 139, 1975.
56. **Aviado, D. M.,** *The Lung Circulation,* Pergamon Press, Oxford, 1965, 25–29.

# AUTHOR INDEX

## A



## B



## C



## D



## E

## F



## G



## H



## I



## J



## K

## L



## M



## N



## O

## U



## V



## W



## X



## Y



## Z

# SUBJECT INDEX

## D



## E

## F



## G



## H

## N



## O

## P



## R



## S

## T



## U



## V



## X

# TITLES OF INTEREST

## CRC PRESS HANDBOOKS AND MANUALS:

CRC HANDBOOK of CHEMISTRY and PHYSICS, 57th Edition
Edited by **Robert C. Weast, Ph.D.,** Consolidated Natural Gas Co., Inc.

CRC HANDBOOK of CHROMATOGRAPHY
Edited by **Gunter Zweig,** Environmental Protection Agency, and **Joseph Sherma, Ph.D.,** Lafayette College.

CRC HANDBOOK SERIES in CLINICAL LABORATORY SCIENCE
Chief Executive Editor, **David Seligson, Sc.D., M.D.,** Yale University School of Medicine.

METHODOLOGY for ANALYTICAL TOXICOLOGY
Edited by **Irving Sunshine, Ph.D.,** Cuyahoga County Coroner's Office (Ohio).

## CRC PRESS "UNISCIENCE"™ TITLES:

AIR POLLUTION from PESTICIDES and AGRICULTURAL PROCESSES
Edited by **Robert E. Lee, Jr., M.S., Ph.D.,** Environmental Protection Agency.

CARCINOGENESIS TESTING of CHEMICALS
Edited by **Leon Golberg, M.S., B. Chir., D.Sc., D. Phil., F.R.C. Path.**

DRUGS as TERATOGENS
By **James L. Schardein, B.A., M.S.,** Parke, Davis and Company.

MASS SPECTROMETRY of PESTICIDES and POLLUTANTS
By **Stephen Safe, B.Sc., M.Sc., D. Phil.,** and **Otto Hutzinger, Ing. Chem., M.Sc., Ph.D.,** both Scientists at the National Research Council of Canada.

## CRC CRITICAL REVIEWS™ JOURNALS:

CRC CRITICAL REVIEWS™ in TOXICOLOGY
Edited by **Leon Golberg, M.D., B.Chir., D.Sc., D. Phil., F.R.C. Path.,** Chemical Industry Institute of Toxicology.
Direct inquiries to CRC Press, Inc.